T0127811

Amphibian Survey and Monitoring Handbook

Amphibian Survey and Monitoring Handbook

John W. Wilkinson

Pelagic Publishing | www.pelagicpublishing.com

Published by Pelagic Publishing
www.pelagicpublishing.com
PO Box 725, Exeter EX1 9QU, UK

Amphibian Survey and Monitoring Handbook
ISBN 978-1-78427-003-2 (Pbk)
ISBN 978-1-78427-004-9 (Hbk)
ISBN 978-1-78427-005-6 (ePub)
ISBN 978-1-78427-006-3 (Mobi)
ISBN 978-1-78427-074-2 (PDF)

British Library Cataloguing in Publication Data
A catalogue record for this book is available from the British Library.

Cover images, top left: parsley frog (*Pelodytes punctatus*), Mark Barber; mid left: dip-netting, John W. Wilkinson; bottom left: western common toad (*Bufo spinosus*), John W. Wilkinson; top right: ©iStock.com/Kalawin; centre: Morogoro tree toad (*Nectophrynoides viviparus*), John W. Wilkinson; centre bottom: palmate newt (*Lissotriton helveticus*), John W. Wilkinson; bottom right: fire salamander (*Salamandra salamandra*), John W. Wilkinson.

Typeset by Saxon Graphics Ltd, Derby

For my Mom, Wendy, who became very good at rearing *Dendrobates truncatus* and could find room for 5,000 toad tadpoles at only a moment's notice …

Contents

List of figures

List of tables

Foreword

A gleam on water. Light illuminates the surface to reveal two long strings of dots like necklaces of black pearls. Under a hefty stone on the gently sloping bank, a natterjack huddles down to get warmth from the surrounding sand. Beyond the bank, females have burrowed below the sparse turf awaiting the call to spawn. Away on pastoral farmland, tadpoles of the common toad have struggled from their spawn entwined around vegetation in deep pond water.

You don't need to go far to see such engaging creatures if you only stop to look, wait by a pond and see what happens. Return after dark with a good torch and see newts courting at the margins and, with luck, the great crested newt in full display: a waggling tail, a silver flash, a crest erect. But such fascinating creatures face a difficult future: humans drain the land, fill in ponds, build houses and intensify agriculture.

Bodies like Amphibian and Reptile Conservation aim to conserve amphibians not just for their own sake but because they are part of the intricate web of biodiversity, in which all creatures play a part and enrich the quality of our lives, if only briefly.

You can become involved and help amphibians. John Wilkinson details how to plan survey work, record and monitor the presence of species that will yield the information vital for conservation policy and practice. John's lifelong experience and enthusiasm for amphibians shines through his book. So, read, enjoy, survey and thereby help give our amphibians a more secure future.

John Buckley
Amphibian Conservation Officer
Amphibian and Reptile Conservation
Boscombe 2015

Preface

By far the most effective way of conducting an amphibian survey would be to know from the start exactly what data you were going to generate and how you were going to use those data (having read every relevant publication and designed your survey protocols accordingly). Few of us, however, have that luxury for most of the time. The aim of this book is to set out the considerations and techniques needed for conducting amphibian surveys so that you will obtain data you can use. I have also used published amphibian survey studies as guiding examples and suggested further reading. With a little forethought, you can make your amphibian survey interesting, useful and effective in achieving its goals.

What is covered?

The *Amphibian Survey and Monitoring Handbook* is divided into three main chapters: covering *before*, *during* and *after* your survey. What questions do you need to ask before you start? How will you carry out the survey and what equipment will be needed? Having completed your survey, how will you arrange your data and what is the best way to tell everyone else what you've found out?

Having read this far you have probably started to ask some of these questions: good! Use the sections of this book to make checklists of the resources and techniques you will use throughout your survey. When planning any kind of study or survey project, I start with a long list of notes and things to do, some of which need to be ticked off near the start and others of which can be dealt with as the survey progresses. As you become familiar with surveying amphibians (or if you already have some experience), your lists become shorter as you acquire the skills and equipment required. Good planning (and some flexibility!) goes a long way towards making your amphibian survey successful.

As well as actual survey techniques, I have also included a section on *handling* amphibians. Believe it or not, I have in the past been asked to advise on ambitious and well-planned surveys of amphibians in remote parts of the world, by people who have never actually held a frog!

What is not covered?

This book does not cover species identification, though the different groups of amphibians are discussed in the first section. It is, however, very important to have a good idea of the range of species you are likely to encounter during your survey, so refer to one of the many excellent national and regional amphibian guides now available (see Chapter 5). Also, though I haven't covered in detail all possible studies involving amphibians, some of these are discussed in Chapter 4. A huge variety of investigations can be carried out, and perhaps your amphibian surveys will inspire you to look more deeply into the biology of these intriguing animals.

And finally...

Most of my amphibian research and surveying has been carried out in the British Isles, where I live and work, so I have drawn on these experiences to populate the chapters of this book. I have, however, tried to include examples and illustrations from different parts of the world to illustrate important points. Many thousands of people around the world now conduct amphibian surveys, whether out of personal interest or for professional or academic reasons, and find it a fascinating and rewarding activity. I very much hope that you do too.

John W. Wilkinson, Dorset 2015

Acknowledgements

Thanks to Mark Barber, Trevor Beebee, John Buckley, Emma Douglas, Dorothy Driver, Matt Ellerbeck, Tony Flashman, Georgia French, Mark Gardener, Tony Gent, David Gower, Stuart Graham, Peter Hill, Jim Labisko, Brett Lewis, Simon Maddock, Liam Russell, David Sewell, Rick Sharp, Thomas Starnes, Ben Tapley, Moacir Tinoco and David W. Williams, who provided comments, ideas, photographs, information or advice for these pages. Last but not least, I thank Nigel Massen and Thea Watson of Pelagic Publishing for their help and patience! Without the help of all these people, this book would not have been possible.

1. Introducing amphibians

Today's amphibians are descended from the first terrestrial vertebrates that evolved from bony-finned fishes some 360 million years ago. A few tens of millions of years later, early amphibians gave rise to reptiles, and ultimately therefore to both birds and mammals. Your ancestors were amphibians!

The majority of amphibians develop from gelatinous eggs laid in water, which hatch into aquatic larvae (tadpoles or polliwogs) and gradually metamorphose into terrestrial juveniles resembling the adults. Many species, however, have evolved variations on this theme. Some salamanders, for example, reach sexual maturity in their larval form and never attain true adulthood, and some frogs produce eggs within which the tadpole stage takes place, hatching into fully formed froglets. Still others undergo the typical amphibian metamorphosis but remain aquatic throughout their lives. For a fuller overview of the diverse range of amphibian life histories, see Cloudsley-Thompson (1999), Halliday and Adler (2002) or Wells (2007).

1.1 Amphibian diversity

Modern amphibians number more than 7,000 species in three orders (see below). This number is only a fraction of the diversity present when amphibians were the dominant terrestrial vertebrates during the Carboniferous and Permian periods, more than 300 million years ago.

Amphibians are now among the most threatened of vertebrate groups, with more than 30% of amphibian species falling into IUCN threat categories (Stuart *et al.*, 2004). Though some amphibians are declining to extinction – many more than would be without the influence of humans (McCallum, 2007) – the number of known amphibian species is actually rising as a result of taxonomic research and advances in methods used for separating species (see Box 1.1 on page 7).

Many factors are contributing to global amphibian declines, including emerging diseases and pollution, but by far the most serious threats to amphibians are the loss and fragmentation of their habitats. The modern, human-dominated world is a hostile place for many amphibians, which have particular ecological needs, including, for example, the need to migrate between hibernation sites and breeding ponds. Despite this, many amphibian species persist and thrive alongside human activities where the features on which they depend are still present in the landscape. Often, it is habitat fragmentation and the *rate* of landscape-scale change that prevent amphibians from colonizing new habitats, such as urban ornamental and sustainable urban drainage system (SUDS) ponds (e.g. Gledhill and James, 2008). For more information on global amphibian declines, see the reviews by Beebee and Griffiths (2005), McCallum (2007), Hamer and McDonnell (2008) and Blaustein *et al.* (2011).

The fact that many amphibians, even familiar and widespread species such as the common toad (*Bufo bufo*) in Europe, are declining (Carrier and Beebee, 2003) is one of the factors driving the need for effective and informative surveys. Whether conducting a site species inventory or carrying out detailed monitoring of a single population, survey results provide invaluable information that can lead to better conservation decision making. Your surveys can therefore inform positive action that will help prevent further declines in these critical components of our ecosystems.

1.2   Order Anura (frogs and toads)

The anurans are the most diverse and numerous of the amphibians, with more than 6000 species in more than 50 families. The familiar frogs and toads are anurans, though the distinction between the two is an artificial one: anurans that jump and have smooth skin are referred to as 'frogs', and those with warty skin and a tendency to walk or hop are called 'toads'. The midwife toads (genus *Alytes*) and fire-bellied toads (family Bombinatoridae) are more closely related to frogs than are the poison-dart frogs (Dendrobatidae) and treefrogs (Hylidae), which are allied to the true toads (Bufonidae).

Anurans lack tails and usually have a moist, scale-less skin that can be smooth or glandular and warty. Most species are also characterized by their prominent eyes, used for locating moving prey, and many species produce distinctive calls, most often as a way for the males to attract a mate. They breed, typically, in water (see Fig. 1.1), using external fertilization and laying eggs singly, in clumps or in strings, in a variety of aquatic situations from small, temporary puddles to large, permanent lakes. The jelly-covered eggs hatch into swimming larvae (tadpoles or polliwogs) that develop over days, weeks or (sometimes) years into smaller versions of their parents. Some of the many exceptions include the viviparous *Nectophrynoides* (see Fig. 1.2) and the Neotropical genus *Eleutherodactylus*. The eggs of the latter undergo direct development, hatching into fully formed froglets. Some species of anuran (e.g. the red-eyed treefrog, *Agalychnis callidryas*) deposit eggs on leaves over ponds or in other situations where the tadpoles can

Figure 1.1 The archetypal anurans are frogs like this Epirus water frog (*Pelophylax epeiroticus*). This species, associated with a variety of water bodies in the lowlands of western Greece and southern Albania, is classed as vulnerable (VU) by the IUCN because of threats to its habitats and commercial collection for human consumption. Small clumps of eggs are laid amongst aquatic vegetation (David W. Williams).

Figure 1.2 Fairly typical (in shape at least) for an anuran is the Morogoro tree toad (*Nectophrynoides viviparus*) from the Eastern Arc Mountains of Tanzania. Unlike the majority of anurans, however, it is completely terrestrial; fertilization is internal and the tadpoles develop within the female until being born as fully formed toadlets. *Nectophrynoides viviparus* is another vulnerable (VU) species because of its severely fragmented habitats, which are being further reduced by farming and other human activities (John W. Wilkinson).

wriggle into the water when they hatch, reducing the risk of their eggs being eaten by predators. Other species, including members of the Dendrobatidae, reduce this risk by practising advanced parental care. Eggs are cared for and kept moist by one or both of the parents and the tadpoles are carried to suitable puddles or reservoirs of water in epiphytic plants to develop in relative safety. Some dendrobatids even produce unfertilized eggs that provide a regular source of nourishment for the growing tadpoles.

1.3 Order Caudata (newts and salamanders)

The order Caudata contains more than 600 species in 10 families. Though there are many fewer species than in the Anura, caudates nevertheless display a remarkably diverse range of life histories. Caudata includes the pond-breeding newts (see Fig. 1.3) of Europe, Asia and North America (family Salamandridae) but, again, the distinction between newts and salamanders is an artificial one based on their life histories. Many salamanders are more terrestrial than newts but equally others, such as the axolotl of Mexico (*Ambystoma mexicanum*) and members of the family Proteidae (including the olm, *Proteus anguinus*, of Europe), are paedomorphic, never leaving the water and never usually developing into

Figure 1.3 The palmate newt (*Lissotriton helveticus*) is a typical pond-breeding newt. Eggs are laid singly, wrapped carefully in the leaves of aquatic plants, and the larvae develop in ponds, but most adults spend summer and autumn, and hibernate, in damp places on land. The species remains common in many of its western European (often upland or acidic) habitats so is classed as least concern (LC) by the IUCN. Despite this, some populations (e.g. in The Netherlands and Spain) are considered threatened (John W. Wilkinson).

Figure 1.4 *Plethodon cinereus* is a widespread and variable lungless salamander from eastern North America. Females guard their terrestrial eggs, avoiding predation and dehydration until they hatch into fully terrestrial juveniles. This appears to be a successful strategy for salamanders, as the species can occur at high density and is regarded as least concern (LC) by the IUCN (Matt Ellerbeck/ http://www.savethesalamanders.com).

an adult form. They retain their larval characteristics and breed, effectively, as tadpoles.

More than half of the species of caudate, however, belong to the family Plethodontidae. This family has a few species in Europe and many in the Americas, and is the only caudate family with many species in the Southern Hemisphere. The majority of this family are terrestrial (see Fig. 1.4), though some live aquatically, and quite a few small, Neotropical species have evolved to be arboreal, using direct development of their eggs to conduct their entire, tiny lives above ground. Plethodontid salamanders are lungless, obtaining all the oxygen they need through their skin.

Unlike anurans, all caudates have tails and a low-slung, lizard-like body pattern. Fertilization is predominantly internal, being facilitated by a spermatophore (packet of sperm) that the female picks up. Many caudates have an intricate and complex courtship ritual that is unique to their species. Eggs can be single or laid in groups/clumps and may be cared for by one or both parents. A few species, like the fire salamander (*Salamandra salamandra*), retain their eggs and give birth to tadpoles or even completely metamorphosed terrestrial baby salamanders. Caudate tadpoles typically possess branched, prominent external gills (these are internal or less obvious in anuran tadpoles). They undergo gradual metamorphosis into adult form in the same way as anurans.

1.4 Order Gymnophonia (caecilians)

The least numerous and least known of the amphibians, the order Gymnophonia currently contains around 200 species in 10 families. Part of the reason for this group being relatively poorly known is that most species are burrowers in leaf litter and moist soil, though some are aquatic. They can therefore be difficult animals to survey and study effectively, and detailed information about their distribution and fascinating ecology is only just becoming known. Caecilians are restricted to the tropics.

All caecilians share the same basic body pattern (Fig. 1.5): an elongated, limbless and virtually tail-less body like that of a large earthworm (for which

Figure 1.5 *Ichthyophis beddomei* is a caecilian endemic to the Western Ghats of India. One of this specimen's subocular tentacles is visible just in front of and below the reduced eye. Though primarily a burrower, the species is also capable of swimming and females lay clutches of eggs, which they care for, near water. The larvae are aquatic initially but become burrowers after just a few days. The species is classed as least concern (LC) by the IUCN. More research will elaborate any threats to its populations (Ben Tapley).

they are sometimes mistaken). The eyes are tiny and indistinct. Their skins are smooth but appear scaly in species that have segmented rings (annuli) around their bodies. All the amphibian groups have extraordinary characteristics and caecilians are no exception. They have (uniquely) evolved a small, mobile tentacle below each eye. These retractable tentacles aid in sensing prey items and mates in the caecilians' underground world where good eyesight would be useless. Additionally, though primitive and simplistic in appearance, they utilize different reproductive strategies. Fertilization is internal, with about half of all species producing eggs that hatch into gilled larvae, and the others giving birth to young that have developed within the female. Courtship in this group is very poorly known.

Some female caecilians, depending on the species, produce a kind of internal milk to feed their developing embryos, and some, like *Boulengerula taitanus* of Kenya, even grow and shed special layers of skin for their hatched larvae to eat. The larvae rasp this skin off their mother's body with special baby teeth (see Kupfer *et al.*, 2006). The resemblance to earthworms is therefore entirely coincidental!

BOX 1.1 AN AMPHIBIAN BY ANY OTHER NAME …

Taxonomists often disagree about how different species of amphibian are related to one another, and change scientific names to reflect the latest thinking. The European newts, for example, all used to be placed in the genus *Triturus*, but this is now reserved for some of the larger species. Closely related smaller newts are now placed in the genus *Lissotriton*. The alpine newt, however, is intermediate between these two genera and so was given a genus of its own (*Mesotriton* = middle newt). So far so good, but then it was realized that the alpine newt had already been given a unique genus name back in 1801 and, under the rules of zoological nomenclature, this has precedence. So the alpine newt is now *Ichthyosaura* (= fish-lizard), as it was when it was first given its own genus.

Is this important to you? Well, possibly not, but it *is* important to know which species you are surveying for, because differences in species' ecology influence how and where you study them. Textbooks and field guides are sometimes behind the times when it comes to scientific names, and that includes times when species are split into two or more other species that may look very similar. So, use modern identification guides when you can (see Chapter 5) and, if these aren't available for your survey area, record when and where you did your surveys very carefully, taking photos if necessary to carry out identification later on. The conservation status of

Continued

species changes too, and your survey results might one day inform the conservation of amphibians that were common when you did the survey. Following these steps will help ensure your survey data remain useful even many years into the future, when the names might well have been changed again (or changed back to what they started as!).

Recent changes to amphibian classification are discussed by Frost *et al.* (2006, a comprehensive paper that some have found controversial) and Speybroeck *et al.* (2010, which updates the list of species found in Europe and their names). The most recent research is already suggesting further changes …

2. Before you start surveying

2.1 Types of survey

Defining the type of survey you are intending to carry out will help make sure you have everything in place to make it a success, as well as avoiding the mistake of having unrealistic expectations about what the results can show you. Broadly, and irrespective of which survey methods are used, amphibian surveys can be divided into three categories.

- *Census or inventory surveys:* these surveys are usually confined to a pre-defined area (e.g. a particular pond, protected area or region) and can reveal the presence/absence of amphibian species there and their relative distributions within that area. Inventory surveys can also be used comparatively to inform ecological questions such as *How do the amphibian species present in Area A differ from those present in Area B?* when combined with other information specific to the areas being studied (e.g. pond size and depth, surrounding vegetation types, geology). Inventory surveys are generally intended as one-off exercises, though careful data collection will mean they can inform more extensive surveys and be referred to for comparisons over time (e.g. if a grassland area gradually becomes woodland, the composition of the amphibian species present may change). The potential for future inventories of the area in question may be unknown, but any inventory survey should be carried out with the possibility in mind that someone will refer back to it in future.
- *Surveillance:* this is a programme of repeated surveys intended from the outset to detect changes over time. The intention might be simply to check that the same species previously recorded in an area continue to be present, or to assess the effects of habitat works or landscape changes: *How has the construction of a new housing estate affected the species composition of amphibians breeding at Pond X?, Are the most sensitive species still present after five years?,* etc. Effective surveillance will usually include information additional to simple species presence/absence, such as an estimate of population size/s,

and each survey *must* include some comparative information to be useful. It is interesting (though saddening) when surveillance shows a protected species is no longer using a pond five years after a road was built next to it ... but what is the actual change that has caused the loss of that species? Changes causing observable effects on amphibians can be many and varied; in the example of the construction of a new road, the changes could be loss of terrestrial habitat, fragmentation of the habitat, harmful runoff from the road's surface, or perhaps simply that amphibians are being killed by traffic on their way to the pond. It is often the case that such factors combine to affect amphibians, so it's useful (but sometimes difficult) to predict and record systematically (i.e. in exactly the same way each time your survey is repeated) a variety of related information (see Section 3.3).

If population size information is included in surveillance, it can be used to detect trends. It is essential to note, however, that many amphibian populations fluctuate substantially over time and that separating a 'real' decline from natural fluctuations may require many years of data (e.g. Pechmann *et al.*, 1991; Blaustein *et al.*, 2011). For example, a new road might result in the recording of slightly fewer amphibians at a time of population 'boom' but might equally cause local extinction of a species if the road construction happens at a time of low numbers of breeding adults. This problem also emphasizes the importance of systematic long-term surveillance of amphibian populations to the investigation of declines, and as an aid to effective conservation.

- *Monitoring:* this is very similar to surveillance, but with the added dimension of having a target or threshold value that is pre-defined and that informs us about the conservation status of the species involved. Monitoring relative to pre-defined values can help keep track of the status of species in an area almost irrespective of the known tendency of amphibian populations to fluctuate (see above). In reality, few amphibians are well-known enough in any area to establish such target values, but this should be the goal of any long-term monitoring. You may, for example, have long-term data that tell you *Newt Species Q* populations in *State Park Z* fluctuated between 75 and 250 breeding adults over a 20-year period. So a monitoring target could be to *detect at least 75 breeding adults every year*. Fewer than this number would trigger more intensive surveys and/or an investigation into a possible 'real' decline and subsequent remedial action. Monitoring thresholds can also be defined in other ways, for example *Is there, in any given year, evidence of breeding (eggs detected) of Newt Species Q in at least 12 of the 17 ponds in State Park Z?* Every monitoring target will be specific to species and area, the critical aspect being that values are based on systematic surveys over a sufficient length of time for them to be both robust and useful (see Section 2.2).

As I said, few amphibian populations are well-known enough to establish adequate target values, but monitoring with targets as defined here can still be enshrined in legal instruments. Member states of the European Union (EU), for example, have a list of species (including the great crested newt, *Triturus cristatus*, and natterjack toad, *Epidalea calamita*) for which they have to report on the conservation status every six years. This responsibility comes from the EU Habitats Directive of 1992. The metrics (population, habitat, range and future prospects) are defined, and status is favourable only when all four metrics can be said to exceed Favourable Reference Values (FRVs). Target (FRV) values against which to monitor, however, are currently poorly explored, so Favourable Conservation Status (FCS) is frequently interpreted or quantified differently by individual EU member states! Many countries use the date when the Directive came into force in their country (often 1994) as a reference threshold, and concentrate their reporting on trying to prove that a species' status has not dropped below that level. In most cases, though, there are no data to say that status was 'Favourable' at that point. Current discussions on monitoring and FCS at EU level may nevertheless eventually lead to useful guidance on setting targets that will hopefully be adopted widely.

Perhaps inevitably, many concepts originating within the EU tend towards complication and I don't intend to explore further the EU's take on monitoring. It is useful to be aware, however, that governments do attempt to measure status through monitoring in this way. More information on FCS can be found at http://jncc.defra.gov.uk/PDF/comm02D07.pdf, and the reports on the status of species by each EU country can be seen at http://bd.eionet. europa.eu/article17/index_html/speciesreport. The real value of the concept of FCS, of course, is to define appropriate conservation targets that can relate back to tangible conservation action: *Has the habitat management carried out in State Park Z improved the conservation status of Newt Species Q there?* If we can monitor amphibian populations against targets in a protected area or a defined region, target values can also be aggregated to produce country values and even eventually describe the status of a species across its entire range. So even irregular species inventories of the amphibians breeding in the pond at your local nature reserve can, if conducted systematically, provide valuable evidence on which practical conservation methods can be based in the long term.

2.2 Survey and monitoring programmes

Many different amphibian survey and monitoring programmes are underway in different countries and regions around the world. No two are exactly the same because they are tailored to the needs and expectations of the area in which they take place. Examination of the protocols of these programmes can help you design your amphibian surveys and make sure all the elements you need are included.

- **FrogWatch Western Australia**
  http://museum.wa.gov.au/explore/frogwatch
  Organizers: Alcoa and the Western Australian Museum
  Covers: frog fauna in three regions of the state of Western Australia
  Protocol: species inventory and distribution surveys in the three regions (each of which has a different frog fauna) through public participation, including promotion of frog-friendly garden
  Notes: one of a series of FrogWatch programmes covering different areas of Australia, where the amphibians and their biology are still incompletely known

- **Netherlands Amphibian Monitoring Programme**
  http://www.ravon.nl/En/RAVONWorkinggroupMonitoring/tabid/382/Default.aspx
  Organizers: Reptile, Amphibian and Fish Conservation Netherlands (RAVON)
  Covers: all Netherlands amphibians
  Protocol: defined amphibian sites are monitored by trained volunteers, each site is visited multiple times a year, for several years in a row, and the species present recorded and enumerated
  Notes: enumeration methods are adapted to species' biology and national status; this is a true monitoring programme that generates population trend information

- **National Amphibian and Reptile Recording Scheme (NARRS) (UK)**
  http://www.narrs.org.uk
  Organizers: Amphibian and Reptile Conservation (ARC), with partners including the volunteer network of ARG-UK (Amphibian and Reptile Groups – UK)
  Covers: all amphibians (and reptiles)
  Protocol: 1-km survey squares are assigned to trained volunteer surveyors who visit the pond in the south-westernmost corner of the square four times in a year, using several survey methods, to record the species present and collect habitat information
  Notes: now generates national and regional occupancy rates; currently being expanded to include sites where population size information is recorded annually to examine trends

- **North American Amphibian Monitoring Program (USA)**
  http://www.pwrc.usgs.gov/naamp/
  Organizers: US Geological Survey (USGS)
  Covers: frogs and toads across US states

Protocol: trained volunteers take on pre-defined roadside survey routes with ten 'listening stops' where vocalizing frogs and toads, plus environmental variables, are recorded
Notes: aims to assess anuran population trends at state, regional and larger geographic scales. Volunteers are required to take a 'skills quiz' before participation

These are just a few examples of surveys with the aim of collecting data on amphibian breeding sites or populations: visit their respective websites for more information on protocols, latest results, and of course if you want to take part! Each of the above programmes has adopted a different approach, including training, species inventories, population enumeration and trend detection. You can ensure your own survey is effective by seeing *what has worked for other situations* and according to the resources you have available. Ultimately, your survey's success will depend on its design within the constraints of those resources and on the aims set out for your survey: what are you trying to find out?

2.3 Survey aims and resources

So, will you be carrying out a species inventory of the ponds in your local park or are you embarking on a five-year monitoring project for the amphibians of an entire country? Whether your survey is a student project, designed to inform habitat management, or a long-term, fully funded status assessment, an amphibian survey should always *ask a question*. Some hypothetical questions were discussed in Section 2.1, and the question you ask will inform your aims. Make a checklist of everything you'll need to carry out your survey and write your survey question at the top! An example checklist is given in Box 2.1. In this fictitious survey, the local council wants to know which species of amphibians are present in its municipal park and how it might best be monitored in future. The park's Friends Group have part-funded the project so that a student can be resourced to conduct surveys over a whole breeding season. The main interest is in informing the management of the park, which will have to be planned carefully if protected species are found to be present. If this was a real project it could be a useful and interesting one because there is an earlier survey for guidance and comparison. The student still needs to work out the budget for travel, buy a few needed items and spend time making funnel traps for use in the survey. There may be other things that need writing down and can be added as the survey planning progresses …

BOX 2.1 INITIAL SURVEY CHECKLIST: OAKWOOD POND SURVEY 2015

Survey Question/s:
Confirm spp. present (protected spp.?) and investigate best survey conditions/ times for future monitoring efforts

Species Involved:
Common frog, common toad, unknown newts (2 spp.?)

Timeline:
Early Feb 2015 – research and planning (send survey plan to Rangers)
End Feb–April 2015 – surveys 3× per week? (inc. night survey with Ranger Thursday eves)
By 24th May – records uploaded to Oakwood Recording Scheme, report to Rangers, Council and Friends Group (NB planning meeting 29th May at 10.30, Council Offices)

Permissions:
Access granted (must be accompanied after dark), licence to survey protected species applied for (chase up by 1st Feb)

Personnel Resources:
(2) self plus Oakwood Park Ranger Ms. Jones (available Thursday evenings)

Funding Resources:
Total budget £800 (£350 from County Council, plus £450 grant from Friends of Oakwood Park)

Equipment Needed:
Park Rangers have GPS, dip nets and torches to borrow, need to make bottle traps (18) and provide own transport, buy thermometer and disinfectant, design and print survey forms (at Ranger Office?)

Information Resources:
Have field guide, need to locate 2006 Oakwood Pond Survey Report (by Joe Smith) and research trapping protocols. Check stats for analysis and software availability

Protocol:
Visual survey (day) and netting every Mon or Tues, set traps and torching Thurs evenings, check traps and terrestrial refugia search Fri mornings. Record esp.

spawn clumps/strings (with locations) and newt numbers for population size class estimates. Check reporting format and methods from 2006

Outputs:
(1) spreadsheets for on-line scheme, (2) report highlighting optimal survey times/ conditions, (3) attend planning meeting if requested, (4) prepare funding bid for follow-up survey and comparison with Birchtown area ponds in 2016???

Notes:
Presence of great crested newts will affect Park management. *Ask pub for empty soft drinks bottles!* Check council mileage rate for budget! Do Risk Assessment

Also critical are the resources you'll need. These will usually include time, personnel, funding, equipment and information. All these factors might dictate what methods you use, how much analysis you can carry out, and how your results are presented and disseminated. If you don't have the resources you need to answer your survey question, you might have to modify the question to fit those you do have, for example from comparing amphibian communities in all the ponds in an area to just one pond in each major habitat type.

The methods used for amphibian surveying are discussed in Chapter 3. The ones you intend to use again have implications for your available resources. Funnel trapping, for example, requires two visits (one to set the traps and one to empty them) whereas a night torching session might only need one visit but two people (for safety reasons at least) for several hours plus time to travel to the pond site. So think about what you can realistically achieve with what's available and don't be afraid to make your survey aspirations more modest at the outset. It's better to plan a robust and repeatable survey protocol for which you can meet all the aims than to be over-ambitious and fail. For example, if you want to compare the numbers and species of newts trapped in a pond over a three-month period, but won't have enough money left after six weeks to support travel to the pond, the survey will not achieve its objectives.

So the design of your survey is affected profoundly by available resources. Limited time or money could mean you have to reduce the number of survey sessions but, if you have to reduce the number of sites (e.g. ponds) that you visit, how can you decide which ones to miss out? Unless your objective is purely to inventory as many species as you can, you should avoid simply trying to go to the 'best' sites. This introduces bias to your survey design and, while seeing lots of amphibians might be interesting for you, your results will be less informative over time and less representative of the real situation. For this reason, many long-term survey projects use a protocol that attempts to reduce bias. NARRS amphibian surveyors in Britain (see Section 2.2), for example, are asked to visit the south-westernmost pond in a randomly allocated 1-km grid square. If for any reason they can't survey that particular square, they're told to try the one to the north of it, and then if necessary all the other squares adjacent to it, going clockwise. In practice they don't usually need to resort to trying all of them! If access to the south-westernmost pond can't be obtained, they should try to survey the next nearest, and so on.

Comparison of two or more areas where you only have the resources to carry out limited sampling in each area should also be treated as objectively as possible. Work out the survey effort (number of samples) achievable with the available time and resources and allocate survey sites within each area according to a random formula. This could be as simple as assigning each site a number and asking a friend to choose six (or however many) numbers between one and your highest number. A better way is to use a random number generator (these are available as part of computer programs such as Microsoft Excel® or on the internet). You could also choose to stratify your sample, for example using terrestrial refugia at identical densities at different sites or deciding to survey three randomly chosen ponds in each habitat type or per 10-km square. You may want to carry out a power analysis (e.g. Hayes and Steidl, 1997) to determine whether your sample size is sufficient to answer your survey question. The statistical power of your survey (to demonstrate possible differences e.g. over time or between areas) changes with sample size, so the basic rule is to take as large a sample as you can manage. This increases your chances of samples representing actuality!

For many amphibian surveys, good presentation of the results is more important than statistical analysis but, if the success of your survey project (answering your survey questions) *depends* on statistical analyses, consult other surveys or a good textbook for further advice before adopting protocols you could find difficult to change and that might not help answer your questions! Find published (peer-reviewed) survey studies on http://www.scholar.google.com. Some suitable books are suggested in Sections 4.1 and 5.4. You should already be considering whether you'll need to employ some statistics and, if so, which statistical tests you'll want to use. Analysis is always a job for after the survey (because it's a mistake to make conclusions before you have the whole picture) but do plan your analyses as much as possible. Refer to examples from similar studies and use the advice given in Chapter 4 to plan appropriate analyses and reporting *before* you start collecting data. This approach will help you decide on the information resources you'll need to obtain to complete the survey project effectively.

Within the total time available to you for your survey project, you will also need to build in time and resources for presenting and disseminating the results (see Section 4.2). If you just need to set aside an hour to upload your species records to an on-line recording scheme, that might not be a problem, but if the park ranger needs a report from you for a Biodiversity Partnership meeting the day after you carry out your last torching session, how can you realistically meet that goal? A realistic timeline should also be included in your pre-survey checklist, which will show landowners, lawmakers and funders that you have planned the survey well. This in turn inspires confidence in the results and, where appropriate, increases the chances that your survey results will be used to inform conservation and/or as the basis for other surveys. You should also consider whether copies of your survey report will need to be printed and posted to stakeholders and funders: is there money in your budget for this?

2.4   Collecting survey data

One of the other resources you'll need for your survey is a well-designed form (or forms) to record the data you collect. You can find examples of forms to hold site information and for recording survey data in Section 5.1. The exact layout will depend on what data you're collecting, but for every survey four pieces of information are essential: *what* did you record, *where* did you see it, *when* was it seen and *who* saw it? These four things make up the basis of any valid biological record and are often arranged leaving *what* was seen until last, to allow recording of more information there. Consider Table 2.1, which holds real data from a casual survey I carried out in 2013.

Table 2.1 records the three amphibian species that were present at that location on that date, but a little more information, including the possibility of recording *absences*, makes these records a great deal more useful. Table 2.2 shows the same data recorded more effectively.

Table 2.1 Essential survey information.

Date:	10th June 2013
Location:	Dewlands Common Pond, Verwood
Recorder:	J.W. Wilkinson
Species seen:	toad, smooth newt, palmate newt

Table 2.2 Adding useful information to survey forms.

Date:	10th June 2013	Time:	2–3 p.m.
Location:	Dewlands Common Pond, Verwood	Grid reference:	SU 078085
Recorder:	J.W. Wilkinson	Contact details:	Toad Hall Main Road Verwood Email: toad@example.com

SPECIES	EGGS?	TADS?	JUVS?	ADULTS?	NOTES
Common frog	–	–	–	–	–
Common toad	–	abundant (1,000s)	–	–	throughout pond
Smooth newt	–	–	–	4 m	all west side of pond
Palmate newt	–	–	–	20 m, many f	some fs may have been smooth newts
Great crested newt	–	–	–	–	–
Other? (state)	–	–	–	–	parrot's feather present

With these extra data, you can tell this was a casual survey carried out in an hour one afternoon: useful for comparing with future survey efforts. You also know exactly where the pond is and the surveyor could be contacted (if they were real contact details!) with any questions about the survey. Use an accurate grid reference to record locations where possible. Many surveyors now use global positioning system (GPS) units, which usually provide both a measure of accuracy and coordinates in the prevailing projection system for your location. For Table 2.2, I found the pond on a UK Ordnance Survey (OS) map and used their format to note the pond location. Make a note of the projection system if it's not obvious which one is being used and record the level of accuracy, especially if your GPS tells you it is low (e.g. >5 m), otherwise your recorded location might appear to be another pond or field altogether! Additionally, maps and aerial photos of much of the world are now available on-line, so if you don't have a GPS unit (or if the battery fails at the critical time) you can often find your location that way.

You can also see from Table 2.2 that toad tadpoles were abundant (no adult toads were seen) and, interestingly (going by males alone), palmate newts were about five times more abundant than smooth newts (*Lissotriton vulgaris*). I hadn't seen smooth newts at this pond before so, until I saw the males of that species, I assumed the females I saw were all palmates (the females are difficult to differentiate unless examined closely). These relative abundance data might be useful in the future because the exotic invasive plant *parrot's feather* is in the pond, and that might affect different species differently over time, causing changes in species abundance.

Two other species of widespread amphibian were not recorded on this occasion. I haven't seen either of them at this pond but smooth newts seem to have turned up recently so perhaps other species will too if there are changes to the habitat. The recording of absences can provide interesting comparative data, especially if survey effort is known, as the ecologies of different species can mean there are different chances of detecting their presence. This is discussed further in Chapter 3. Note that Table 2.2 also has space for recording 'other' species. These might include rare, local or uncommon alien species not often encountered, and this means that this particular table could be used on a form to record very basic amphibian survey data at most ponds in Britain (and could also be adapted for other places).

So far so good: but what data are missing that might be needed? Well, I know that this particular pond has good access all round and usually good visibility, so my survey method on this occasion was 'visual search' only. You should, however, normally record all the survey methods used, how much of the pond was surveyed and the number of survey visits made (dates and times) as it's been shown that the use of more methods over more visits increases the chances of detecting all the species present (see Sewell *et al.*, 2010 and Box 3.3). The other element of recording survey effort is to note the conditions (weather, temperature, etc.) under which the survey was carried out. Though amphibians can be detected under imperfect conditions, the chances of detecting all the species present will be lower than when conditions for surveys are best. Recording survey conditions provides surveyors repeating your survey a means of assessing, for example, whether an additional species they found was 'new' or whether conditions were simply unfavourable for the detection of that species when your survey was carried out.

Though maximizing your chances of detecting all the amphibians present in your pond or survey area could be the single most critical thing for your survey, what might make it most useful in the long-term are the comparative data you collect (see Sections 2.1 and 3.3). If you have formulated your survey question/s by now, you'll have some idea of what you want to know and therefore what you need to record. Habitat data are often essential, of course, and ways in which these and other data can be collected are suggested in the following chapters. Probably the best advice I can offer is to track down published examples of similar surveys (or previous surveys of the area you're interested in) and consider whether what has worked before will work for you. This advice also applies to the analysis of your data and how you present it to other people (including future surveyors who will refer back to your reports for the same reason!). As well as libraries and textbooks, use http://scholar.google.co.uk/ and other web resources to search for informative surveys. Try simple combinations of keywords such as 'frog', 'survey' and 'Colorado', or 'amphibian', 'status', 'Italy' and 'methods' to locate relevant publications. With a little practice and experimentation with your search terms you will find specific studies that will provide lots of ideas. The arrangement and presentation of survey data are also discussed in Chapter 4.

The collection of data is usually easiest on a form laid out exactly the way you want to collect the data in the field. I generally use paper forms printed up with blank tables similar to Table 2.2. Other ideas for setting out this information on forms are shown in Section 5.1. It is usually straightforward to transfer data collected in this way to a spreadsheet or database from which you can generate biological records, graphs, charts and reports in order to disseminate your results (see also Section 4.1). You could also design such a form for use on a tablet computer or other mobile device to record survey data. This can save you time in typing up your survey results, but might also result in the loss of a great deal of data if you drop your mobile device into a pond! I have no doubt that mobile apps, and other methods that can be adapted for amphibian survey and monitoring, will become increasingly common but, until waterproof tablets are widely available and affordable, a weatherproof clipboard, paper forms and a pencil remain the most practical method of carrying out amphibian surveys.

2.5 Survey permissions and licences

Obtaining permission to carry out your surveys is always good practice. If your survey area is on private land you will definitely need permission, but even open access areas usually have someone responsible for them, so permission to survey should be sought. In addition, protected or designated areas may be freely accessible but carry a requirement for a survey permit from the managing authority (such as National Nature Reserves in England, where each survey usually requires permission from the statutory agency Natural England). Apart from not wanting to be challenged or arrested, another good reason for talking to the landowner or manager prior to your survey is that they might be very interested in the results, and you will help instil a sense of pride in 'their' wildlife if you offer to send them a copy of your survey results! If, for whatever reason, survey permission is refused, you will of course need to do some re-planning. Are there alternative sites you can visit that will help answer your survey question?

Whereas seeking permission is always good practice and good advice, the need for a licence varies depending on whereabouts your survey is located and which species are involved. Just within the British Isles, for example, a licence will always be needed to capture, handle or disturb (including surveying) the amphibians protected at a European level (currently the great crested newt, natterjack toad and pool frog, *Pelophylax lessonae*) in England, Scotland and Wales, but is not needed for other species. The requirement for a licence to disturb includes all stages (eggs, tadpoles, juveniles and adults), as well as places of shelter. In contrast, all amphibians are protected in the Isle of Man and in Jersey, but none are protected in Guernsey! In Northern Ireland, surveys for the smooth newt require a licence because of the particular legislation there covering trapping methods. For further information on licensing and legal protection relating to amphibians in the various jurisdictions of the British Isles, see http://www.arc-trust.org.

Always refer to the regulations pertaining to the areas, regions or countries where you are surveying in case there are rules or laws about certain species. Permits to handle threatened species in the USA, for example, are usually issued at state level so you may need more than one if your survey crosses state or national borders. This advice applies even if there are protected species you *might* encounter: will you have to stop surveying altogether if you find species for which disturbance is illegal and you have to wait for a licence? Information on *some* national protective legislation is given at http://www.iucnredlist.org/ – search for the species in which you're interested – but always do more research on what protective legislation will mean in practice for your planned survey.

2.6   Health and safety, and biosecurity

Most companies, charities, universities and other organizations now have a Health and Safety (H&S) policy, intended to keep you safe during your activities on their behalf and to comply with relevant legislation. You should make sure you are aware of such policies during the planning stages of your survey project and ensure you are compliant with them. This is an unexciting but important part of carrying out your survey responsibly: you won't achieve your survey objectives if you end up in hospital!

A usual part of H&S policy is the requirement to carry out a Risk Assessment. This can be daunting if you haven't done one before but it is just the process of thinking about what hazards you may face during survey activities that leads to you actually considering the real risks. Carry out a preliminary site visit in good daylight (when you don't do any surveying, see Section 3.1.1) to record site features and potential hazards before you start your survey programme proper. Hazards can involve anything from ponds to hostile locals, and everything in between. Some common risks associated with amphibian fieldwork are given in Section 5.2, as well as a method of assessing the severity of the risks you might face. Most importantly, your Risk Assessment should list the 'control measures' associated with each risk: how will you minimize the risk of falling into a pond and what action will be taken if you do? How will you prevent your mobile phone from getting wet in case you need to call for assistance?! Organizations where staff or volunteers conduct fieldwork regularly will usually have example Risk Assessments listing appropriate control measures that can guide you when preparing your own.

Some risks can be minimized by having two or more people carrying out surveys together. Many organizations, however, also recognize that this is not always practical and have a lone working policy or procedure to cover this. Often, this involves having a 'buddy' who you will call by mobile phone if you get into difficulty and who can get help to you if you don't check in with them at an agreed time. This is a particularly good idea when surveying at night (whether alone or with someone else). So your buddy will need to know your phone number/s, and those of anybody with you, exactly where you will be going, when you intend to be finished by and how to direct the emergency services to find you if the need arises. This should include place names, grid references and/or postal codes if available, points of access to sites (locations of gates, etc.), as well as possibly the names, addresses and contact details of the landowners. In remote locations, it can also be useful to discover and write down the name of the nearest building to your survey site that has a landline phone. All this information should be part of your Risk Assessment and should be carried with you during surveying, and an identical copy should be left with your buddy.

I have encountered plenty of hostile locals (including cows!) while amphibian surveying, and was once stuck fast up to my waist in Welsh mud while carrying out a survey alone. So it's always best to have rescue plans in place, just in case the need arises. It is also a very good idea to always carry some basic safety equipment, such as:

- a mobile phone (fully charged, and in a waterproof bag is a good idea)
- a torch (I always take a spare torch and spare batteries for night surveys)
- a throwline (for any activity near ponds)
- a First Aid kit (including plasters to cover cuts)
- hand cleanser (alcohol-based, which evaporates after use).

Many amphibian surveys will of course involve contact with pond water and it's easiest to assume that every pond or stream is a potential source of disease. There are several pictures in this book showing amphibians being handled without gloves, but some people use them routinely to minimize any disease risk to themselves. Whatever your choice in this respect, be scrupulous about hygiene once your survey session is over and never immerse open cuts or wounds in pond water, or allow them to contact any amphibians you are handling (see Section

2.7). A few people are sensitive to the mucous skin secretions of amphibians, so for them gloves will be essential. For studies or surveys that involve handling many amphibians in one session, it can be useful to wear gloves so that they can be changed regularly during the course of the survey, minimizing any build-up of mucous, mud, etc. on your hands and helping prevent the transfer of these materials between the animals you are handling. Note also that some H&S policies will require you to wear gloves during fieldwork.

Several types of thin 'laboratory' gloves are available that are suitable for use in amphibian survey work (see Fig. 2.1). These are made usually from latex, vinyl or similar materials such as nitrile (note that some people have an allergy to latex). Additionally, research has shown that using either bare hands or nitrile gloves (as opposed to latex or vinyl) can be beneficial in reducing survival of the amphibian chytrid disease on the hands (see Mendez *et al.*, 2008), and therefore reducing the chance of spreading it. At sites where amphibian diseases are known or suspected to be present, however, gloves should be worn and (ideally) changed after handling each amphibian.

Measures for your own health and safety can also count towards the biosecurity measures for your survey. You should always consider any impact on the pond resulting from your survey activities, and this certainly includes minimizing the potential to spread diseases and invasive species (e.g. exotic

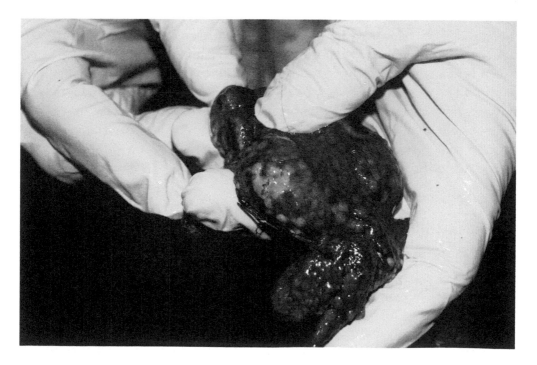

Figure 2.1 Nitrile gloves in use during a survey. This toad has an unusual orange marking (John W. Wilkinson).

pond plants) to other sites. Other possible impacts are discussed under the survey methods described in Chapter 3. To reduce the possibility of spreading diseases and other undesirable organisms, however, always return amphibians you've examined to the same place you found them and clean all equipment between surveys of ponds more than 1 km apart. For survey projects involving ponds less than 1 km apart there is no need to do this because the amphibians themselves, and other aquatic organisms (or ducks, etc.), will likely be able to move between the ponds anyway. Equipment includes nets, buckets, your boots, thermometers and other measuring equipment, and the clothing you're wearing. It will also include your vehicle's tyres if you drive your vehicle up to the pond. Be completely scrupulous with your cleaning if diseases or invasive exotic pond plants are recorded from any of your survey locations.

There are several options for cleaning equipment and vehicle tyres. Firstly, however, all debris, mud, algae, plant material, etc., should be removed before you leave your survey site. Thorough cleaning can then be carried out back at base using either (i) a 70% ethanol solution, (ii) a 4% bleach solution or (iii) a veterinary disinfectant. I have cleaned much survey equipment using a 4% bleach spray solution bought from supermarkets, and this is often easiest in cases where you need to clean your boots, for example, between sites, but bleach will reduce the useable life of pond nets and other equipment so may not be best (or cost-effective) in the long term. Fortunately, these days, veterinary disinfectants such as Virkon (this and others are available via the internet) can be acquired in convenient powder or tablet forms that enable you to make up useful amounts of solution to disinfect just as much equipment as you need. Check what is available locally to you, check its specified disinfectant ability and how it should be used (reputable suppliers provide this information on their internet adverts) and choose accordingly. You should state that you will take/have taken responsible biosecurity steps in both survey plans and reports. This will again inspire confidence in your surveys and demonstrate that you are serious about responsible and robust results.

2.7 Handling amphibians

Many surveys can be carried out adequately without handling the amphibians involved at all. Night torch counts, for example, require mainly counting and recording the amphibians seen: so long as all the species present can be identified in your torch beam. Other surveys will of course benefit from or require at least some amphibian handling, such as when data on their sex or morphometrics are required, or to examine or photograph them for identification purposes. Amphibians you have trapped may also need to be handled carefully prior to their release. If there's a chance you'll handle any amphibians, make sure your hands are clean, and that there is no suntan lotion or insect repellent on them.

In many cases it is most time-efficient, and therefore better for the amphibians, to have two people present: one to do the holding and the other to examine, take photos and record data. If you are unused to handling amphibians, my advice is to be firm but gentle. Many amphibians won't struggle or wriggle excessively if held in this way, but there are always some (especially pond frogs and treefrogs with muscular hind legs) that will! In the examples illustrated, the amphibians are being handled to determine their sex (Figs 2.2 and 2.3) or as part of doctoral research projects (Figs 2.4 and 2.5).

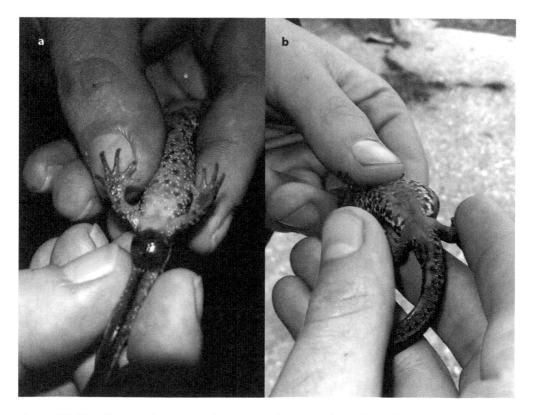

Figure 2.2 Handling small caudates. Some amphibians, such as these *Triturus pygmaeus*, have species-specific sexual characteristics: in the case of breeding *T. pygmaeus* the male has an obvious crest. Before coming fully into breeding condition, however, sexual characteristics are not always fully developed and can be difficult to be sure of. Here, the much larger and more bulbous cloaca of the male (a) is seen easily on brief examination, as is the much smaller cloaca of the female (b). The animals are being supported confidently and are not struggling. In this case, a few moments of indignity is all that is needed to determine sex before the animal's release. Though cloaca size is generally a good indicator of sex in many newts and other caudates, refer to modern identification guides for species-specific information (Mark Barber and John W. Wilkinson).

Figure 2.3 Handling a large anuran. This large (90 mm) *Bufo spinosus* was a particularly lively individual and is being held firmly around the waist/hips, without constriction, so that it has less chance of struggling and crashing to the ground. Smaller amphibians can also be restrained briefly in this way, avoiding damage to their extremities. This toad can be identified as a male by its extremely muscular ('Popeye'-like) front legs: females tend to have slimmer front legs with a more uniform thickness along their entire length. Some male anurans also develop prominent nuptial pads on their inner fingers (used for gripping the female during mating; see inset, also *B. spinosus*) and others have darkened throats or even obvious vocal sacs that can reveal their sex. Again, such sexual characteristics are species-specific so always refer to good contemporary identification guides (John W. Wilkinson).

Smaller amphibians, including juveniles, are inevitably more difficult to handle safely. Tiny toes, tails and organs can easily be damaged by boisterous handling so, if small amphibians have to be restrained, do so by holding the limb girdle areas (the shoulders and hips) and avoid the extremities and soft underparts. Enclosing a small frog or newt in your cupped hands can also help to calm it before being examined. It is often better, however, to examine small amphibians in a tray lined with damp moss or tissue paper, or, in the case of aquatic specimens, a small, clear plastic container containing a few centimetres of water can be used, which also allows animals to be examined from underneath while they are swimming around. Tadpoles and fully aquatic species should preferably always be examined in this way.

Amphibians often urinate when being handled: this can be enough to deter a potential predator (which of course is what the amphibian assumes you are when

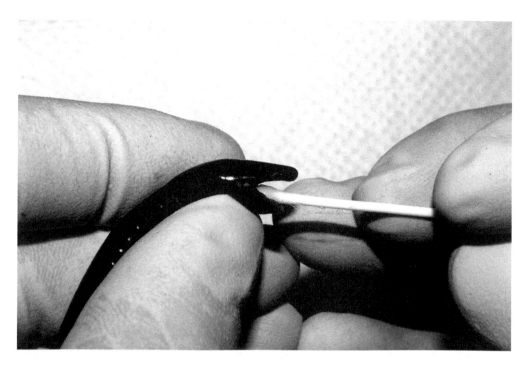

Figure 2.4 This caecilian (*Grandisonia sechellensis*) has been anaesthetized to have a buccal swab taken as part of a genetics study on Seychelles herpetofauna. Studies on caecilians are relatively uncommon but these animals are notoriously slippery customers. If handling this group of amphibians as part of a survey, you'll need to grab them quickly but carefully, then take care that their wriggling and moist skin doesn't cause them to fall to the ground: of course they have no legs with which to break their fall! (David Gower).

you pick it up). In any case the process of being handled can be dehydrating, so any amphibian being held for more than a minute or two should be dampened with a splash of water (along with your hands) to reduce stress and dehydration. Additionally, it is rarely necessary to be standing up while examining your captures; kneeling or crouching will create less panic in any amphibian you're holding as it won't feel as if it's in the sky. Retain captured amphibians in a bucket or other container with a little water or damp vegetation while they are waiting to be sexed, identified, etc., and hold the one you're dealing with over a second, similar container. If it then wriggles free it has less chance of coming to harm and/or escaping. Amphibians should only be restrained with force if they are in danger of leaping out of your hands to their doom!

Finally, if you are fascinated by amphibians but nervous of handling them, get some experience before embarking on your survey. Volunteer to help with another survey or find somebody nearby who has captive amphibians that you can practise holding. There is no substitute for the confidence you will gain from a little first-hand experience, and this will contribute to both the success of your survey and the continued welfare of the amphibians in which you are interested.

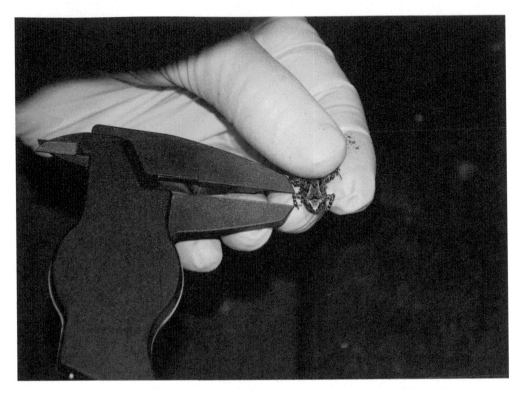

Figure 2.5 Handling a small amphibian for taking measurements. This small *Sooglossus* frog is having its foreleg length measured with callipers for a morphometric study. The animal is being restrained gently by the hip girdle to avoid damage to small limbs and organs. Ways of holding different species and sizes of amphibian become second nature with practice and the key to carrying out such studies safely and effectively is experience. Callipers, balances and other equipment suitable for use in morphometric studies are available from suppliers suggested in Section 5.5 (Jim Labisko).

3. During your survey: amphibian survey methods

Many amphibians rely on water bodies for breeding, so aquatic surveys are often the most resource-effective means of carrying out a survey. This is true particularly for temperate species that either breed explosively in the spring or spend several months living aquatically. If you're confident in identifying spawn and tadpoles (or willing to try), aquatic surveys for even explosive breeders can be extended for sometimes many weeks after breeding adults have left the water. Nevertheless, terrestrial survey techniques are also useful for non-breeding amphibians and for those species that are fully or mostly terrestrial. This part of the book is therefore divided into aquatic and terrestrial survey techniques, though of course some survey projects will use both, either concurrently or at different times of the year, and others don't fall neatly into either category. Combinations of techniques can be most effective at detecting all species present and in providing the most useful information (e.g. Box 3.3). Advice on recording habitat and other comparative information is given in Section 3.3.

3.1 Amphibian surveys in aquatic habitats

3.1.1 Preliminary visits for aquatic surveys

As discussed in Section 2.6, a preliminary visit is useful in planning your surveys and essential for familiarizing yourself with the site to comply with H&S best practice. It is also good practice to sketch your survey sites to remind you of any specific hazards and refresh your memory about site characteristics. Using a sketch means you can be selective about which features are noted but digital photos can be very useful too, especially for recording unique characteristics of the survey site. Site sketches and/or photos can be included in reports to form part of the comparative information available to future surveyors. You can just show basic information and locations of key features, or include anything that might be useful later – during analysis and reporting – such as major habitat types. Either way, sketches should include a scale and indicate north. Features can be added to the sketch as your surveys progress.

The inclusion of additional information can of course help address your survey question/s and should be recorded consistently at all sites. As well as habitat and access information, you should also record, as accurately as possible, the locations of any traps you set so that they can be relocated and collected efficiently. This minimizes any chance of harming trapped animals. It can help to have consistent 'survey stations' (points used for several different survey techniques, see Box 3.1), though it's a good idea not to make these hard and fast until you actually start (in case something prevents you using the ones you intended). The outlines of the pond and habitat boundaries can be drawn most accurately using a grid (and you can also refer to on-line maps and photos), which will also help you decide on survey stations and determine characteristics such as pond area. An example of a pond sketch from the Oakwood Park Pond Survey is shown in Fig. 3.1 and an example preliminary site survey form is shown in Section 5.1.

It's useful to make a copy of your site sketch/preliminary site form and take it with you for reference when you start surveying. Any changes that occur on the site during the course of the survey will thus be more obvious to you. Recording fundamental pond and site characteristics before you start your main surveys also means that most of the data you'll need to record later on will be those that vary with each survey (species and numbers seen, survey conditions, methods used, etc.). This can save time when the effort you can expend at each site is limited. However, ponds do dry up, trees fall over, habitat management occurs and access permissions can change! None of these occurrences is necessarily a disaster but substantial changes potentially affecting your results demand that a new, dated, sketch is done. Changes to the pond or survey site should be mentioned in your final report, even if you think they have had no measurable consequences. They will again provide useful information for future surveyors.

Though aquatic amphibian surveys can be carried out without much specialist equipment, a good torch and a good net will improve the success of your survey. Your equipment list for aquatic surveys might therefore include:

- a clipboard (ideally rainproof), pencils and pre-printed survey forms, extra paper or notebook
- waterproof wellington boots and/or waders
- a hand lens
- polarizing sunglasses
- close-focus binoculars
- a camera
- a small aquarium net
- a high-powered torch (see Section 3.1.3) plus separate head torch and spare batteries

- traps (see Section 3.1.4)
- a pond-dipping net (see Section 3.1.5)
- bucket/s
- clear plastic boxes or white plastic trays for examining amphibians
- an identification guide.

Equipment suppliers and a range of regional identification guides are listed in Chapter 5.

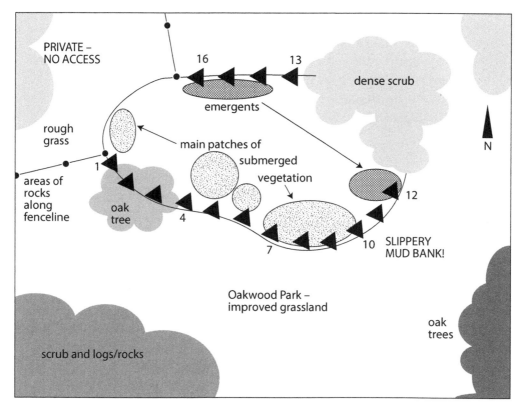

Figure 3.1 Preliminary survey visit sketch of Oakwood Park Pond. Numbered survey stations are shown by black triangles and fall in different habitats around the pond (shaded/open, vegetated/ not vegetated). I've shown much of the information I would usually try and include on a site sketch, though you could add just about anything that might be useful. The grid and scale I'd use to draw the sketch and some of the survey station numbers are omitted for clarity in this example. Even at this level of detail, site sketches can easily become cluttered, so leave out anything irrelevant. You can always add things later and/or do additional sketches showing locations of wood piles, any areas of spawn, etc. From this sketch you could work out the length of the pond's perimeter (about 48 m in this fictitious example), pond area and proportional cover of different habitat types. You can also see that ca. 25% of the perimeter is inaccessible because of lack of access and dense scrub. These site characteristics can be enumerated and recorded alongside your sketch for future reference (see above and Section 5.1) (John W. Wilkinson).

BOX 3.1 FOLLOW THE RULE OF TWO …

Visual searching, night torching, netting and funnel trapping can all be guided by the 'rule of two'. In other words, traps can be spaced 2 m apart, a 2-m arc is a convenient distance through which to sweep a net or shine a torch, and 2-m sections are a good size to search systematically, one after the other, as you make your way around the pond. You can note survey station locations on a sketch as suggested or simply mentally divide up the perimeter of the water body into sections as you survey: a distance of 2 m is about the same as two big paces. As with many techniques, this becomes easier with practice!

Using the rule of two can also help quantify your survey effort and therefore improve the comparability of surveys, for example if you detect three newts per 2-m section at one site and five newts per 2-m section at a second. Your results can of course still be reported as, for example, newts encountered per metre of pond shoreline.

3.1.2 Visual searching

Visual searching is, of course, a more systematic version of 'having a look'. It is nevertheless a valid survey technique and can be standardized by using the rule of two and recording the length and proportion of the total pond perimeter surveyed. Your visual survey really begins as you approach the pond because you might spot and identify amphibians that jump into the water as they sense your approach. Other amphibians could be floating at the pond's surface and swim into hiding as you draw near. Once you've recorded any *obvious* amphibians, eggs or tadpoles in your 2-m section, make a closer examination of weedy or reedy areas, and sections of bank that may slightly overhang the water and therefore provide hiding places. A small (aquarium type) net can be useful to help with this. All visual surveys should, initially at least, be carried out from the bank, to avoid disturbing the pond too much and therefore driving all its amphibians into hiding. Polarizing sunglasses and close-focus binoculars can help you to see and identify amphibians that are underwater or further away from the shore. If arboreal species could be present, don't forget to search reeds or branches overhanging the water.

In locations where you might find breeding newts or other species that affix small eggs to submerged vegetation, a visual search is a good time to look for folded leaves that are hiding newt eggs (Figs 3.2 and 3.3), as well as small clumps of eggs that might be well-hidden. Not all species leave convenient gelatinous clumps of spawn (Fig. 3.4) or long strings of many thousands of eggs where you can easily find them, so as usual refer to good local guidebooks and check the preferred egg-laying sites of the species you might encounter in your survey area.

The total number of egg clumps or spawn strings seen on a visual survey is useful to record; if done accurately it can indicate the number of breeding females at that site (e.g. Beebee and Buckley, 2014). For species that spread their eggs around singly or in small clumps, however, recording numbers of eggs seen is not terribly useful: 400 single newt eggs might equally come from one female newt or from 400 females! When noting the presence of newt eggs in surveys in Britain, I generally just make a distinction between 'a few' and 'many': they can be easier to see if there are lots of females competing for good egg-laying sites, something you'll hopefully confirm, perhaps while torching later on. The same is true for numbers of tadpoles seen (though relative abundance of the different species of tadpoles present in aquatic environments can be a study by itself). A small hand lens can be useful for looking at the differences between eggs and tadpoles of different species.

Figure 3.2 Spotting newt eggs. Several (living and dead) folded leaves revealing the presence of breeding newts can be seen (circled) just under the surface of the water in this picture. On this occasion, fairly large leaves were available, but smaller leaves are also used, as well as just about anything that can be pressed into service (see Fig. 3.3 and Section 3.1.7) (John W. Wilkinson).

Figure 3.3 Newt eggs on folded grass. In ponds where aquatic vegetation is scarce, any suitable substrate can be used for newt egg laying. Here, one or more female newts have folded this submerged blade of grass repeatedly and inside each fold is a single egg. From the egg actually visible on the right-hand side of the picture, you can even identify that at least some of these eggs are those of great crested newts. This species has larger (about 4 mm) and more yellowish eggs than the other newts found in Britain. As a protected species, a licence would be required to unwrap a leaf-fold and check the species' identity in Scotland, England and Wales. Even if this seems necessary and you have obtained the required paperwork, it should still only be done sparingly, as the number of eggs detected has no relationship *whatsoever* to the number of breeding females there. Newt eggs can (and should) be carefully rewrapped after examination and lodged amongst other submerged leaves to give them a good chance of hatching. Also consider that different species of newt (and other amphibians) prefer to lay their eggs at different depths, so don't ignore any folded leaves or leafy areas further below the pond's surface, where the eggs of different species might be lurking, as well as the leaves of terrestrial plants dangling into the water (John W. Wilkinson).

Figure 3.4 (opposite) Spawn clumps of the European common or grass frog (*Rana temporaria*). Amphibians behave differently according to the pond (spawning) conditions available to them. The single spawn clump (a) has been laid amongst aquatic vegetation in a sunny pond, where the warm water will promote embryo development. In (b), however, which was taken at a very shady pond, only this area of the pond is suitable for spawning where it gets some sun for at least part of the day, which will help hatching under relatively cool conditions. At least 90 spawn clumps at different stages of development are visible in (b), meaning the size of the adult frog population breeding there is about 180; *R. temporaria* has a sex ratio of around 1:1 (males: females; Savage, 1961). Many other anurans from around the world will of course select spawning sites according to their own

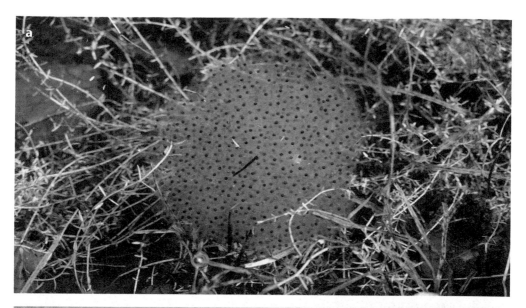

criteria, lower down amongst aquatic vegetation or hidden in the substrate pebbles for example, in some cases to *reduce* exposure to ultraviolet (UV) radiation from strong sunlight that might actually damage the embryos. They may also produce much smaller clumps, strings or even single eggs that are scattered around the pond to spread the risk of predation. For these species, as well as in turbid water bodies with murky water, or where silt can settle on eggs and disguise them, greater visual search effort will be required to detect eggs, and this needs factoring into the time required to complete your survey (David Sewell and David W. Williams).

Once a survey section is complete, walk slowly along the shoreline stopping every 2 m to repeat your search but also scanning ahead as you go. For medium or large ponds, you may want to enter the water (wearing wellington boots or, for deeper ponds, chest waders) to search promising areas of vegetation not easily accessible from the shore. Never enter deep water without a second person present as it can be easy to get stuck or trip, and don't enter the water at all if you've *any* reason to think it won't be safe. Use a long, stout stick or the handle of a dip-net to feel your way around the pond if you can't see the substrate, especially if it's your first time wading in that particular pond. For very large ponds, it can even be useful to use a small boat or dinghy to examine distant areas. Always, however, bear in mind that any disturbance to the pond might spook any animals you could detect more easily later on with a torch or by dip-netting. To start with, the use of binoculars is certainly preferable.

3.1.3 Torching/night surveys

Night surveys using a high-powered torch (Fig. 3.5) can be a very effective method of detecting amphibians in and around a pond. If conducted systematically, torching can be better for gaining comparable population size estimates than other methods such as trapping (e.g. Kröpfli *et al.*, 2010). Responsible torching merely surprises the pond's amphibians for a few seconds and so usually has a low impact on the pond environment and its occupants. I always use my head torch on a low brightness setting when approaching the pond to reduce the warning to any amphibians there that I'm on my way. This also preserves the batteries of your main torch so you can maximize survey time at the pond.

The quality and brightness of the torch you choose for surveying really is important: I have tried several types in efforts to save both weight and money! Modern LED torches have a good battery life but can be expensive and don't seem to penetrate below the water's surface as well as conventional torches. Half-a-million or (better) one-million candlepower torches are easily available but (at the time of writing) can cost more than £100 (US $150) each. They have rechargeable batteries and come generally with a mains-supply charger and/or vehicle charger. If used regularly they are good value but, where they're not, two or more (very much cheaper) DIY store torches with the same candlepower rating can be a better bet and more economical: they are usually lighter but individually don't have the battery life of professional survey torches and their brightness can fade very rapidly.

So, repeat your visual search at night using the best torch or torches you can obtain with the resources available. Sweep the torch through your 2-m survey section in the direction around the pond in which you're surveying (Fig. 3.6).

Figure 3.5 A torchlight amphibian survey. Despite the bright flash needed to take this photo, the high power beam of the torch being used can still be seen illuminating the substrate of this pond, where breeding newts, tadpoles and/ or other amphibians could be detected (Tony Flashman).

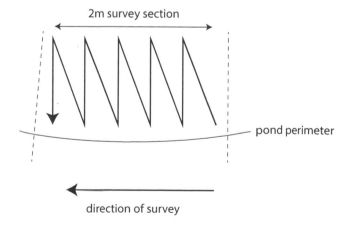

Figure 3.6 Night torching survey method. The zig-zag line shows the path of the torch beam through the survey section. You can extend the distance surveyed towards the centre of the pond for as far as is practical (i.e. if you can still see the substrate or vegetation where you might spot amphibians). Again, don't forget to check for tree frogs on emergent or overhanging vegetation in locations where they could be present (John W. Wilkinson).

The numbers of each different species should be tallied on your survey form as you go: don't rely on your memory and try to write down the totals at the end. Any night surveys of ponds should have a second person present, so that the person not holding the torch can do the form-filling. A clicker-counter can also help you to keep track of the number of amphibians seen, though this is less useful when you're tallying individuals of several species.

Torching is far less effective in rainy or windy weather; it's harder to see the amphibians. Similarly, the method may not yield useful and comparable results in ponds that are consistently murky or that have abundant aquatic vegetation.

Night surveys for calling male anurans (i.e. auditory rather than visual searches) can also be carried out and these are discussed under terrestrial surveys (Section 3.2.6).

3.1.4 Funnel trapping

Various types of aquatic funnel trap have been around for thousands of years, seemingly being invented independently many times over and utilized by the world's waterside cultures to trap fish, eels and many other types of aquatic fauna for food. Fortunately for us, they are also effective for catching aquatic adult amphibians and tadpoles! Though the exact reasons why aquatic animals enter (un-baited) funnel traps are poorly understood, they're probably seeking food and/or shelter in suitable crevices. Funnel traps vary from small, cheap and simple designs that can be made easily at home, to larger, commercial types that require greater resource considerations. This section discusses the types most useful for amphibian surveying.

Bottle traps are the simplest and cheapest type of funnel trap used in amphibian surveys, which makes them very attractive as a survey technique, especially when resources are limited. They can be constructed easily from washed two-litre plastic drinks bottles: cut off the tapered end from such a bottle, invert it and secure in place with a cane (also used to anchor the trap in the pond substrate) at a slight angle; the process is shown in Fig. 3.7. Either clear or opaque (coloured) bottles can be used. The open end of the trap should ideally touch the substrate of the pond and face into open water. Bottle traps tend to work best placed next to edges (e.g. of vegetation clumps, reeds or other features), which help guide any amphibians inside (see Figs 3.8 and 3.9). Sheltered spots are usually better too.

Bottle traps are normally set at dusk and left in place overnight, being checked and emptied early the next morning. The amount of time they can be left safely, however, decreases in warmer weather, sometimes meaning a late night *and* an early start the next day to ensure the safety of any captives.

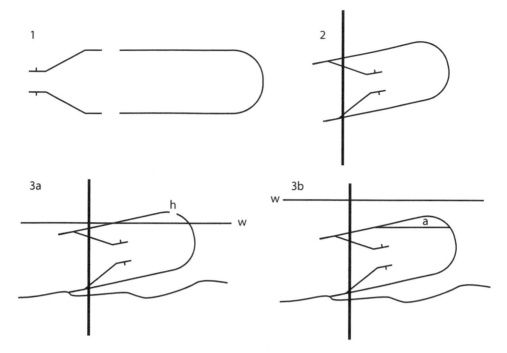

Figure 3.7 Making and setting bottle traps (John W. Wilkinson).

Making the trap
1. Separate the tapered end of a two-litre drinks bottle (using a sharp knife or scissors).
2. Invert the tapered end within the bottle and secure, at a slight angle, using a suitable cane as shown (bamboo is often the best and cheapest type of cane available). If your preliminary survey has shown steep banks, you can adjust the angle of the cane to compensate (by making the angle greater).

Setting the trap with access to air
3a. *Air hole method:* make a small air hole (h) in the highest part of the trap and set it with the hole protruding above the water's surface (w) and secured with the cane to the pond's substrate. This offers the best chance of survival for trapped amphibians, and any other creatures caught.
3b. *Bubble method:* in a minority of cases where water level rises may reasonably be expected, contain a bubble of air (a) in the bottle trap before setting. This can work just as well as (3a) for newts and tadpoles but increases the risk of mortality, especially for accidentally trapped aquatic reptiles and mammals (e.g. snakes and shrews).

Bottle traps have a trap aperture of about 2 cm, making them most useful for small- to medium-sized amphibians, including many newts and most tadpoles. As implied above, however, funnel traps of *any* kind are *indiscriminate*: they'll catch anything that can fit through the opening, including fish, invertebrates, aquatic snakes and small mammals! It is therefore only responsible to provide access to air to maximize survival for *any* animals trapped, whether the intended species or not. This is especially important in warm ponds (including normally cool ponds on warmer days), where the level of dissolved oxygen will be lower,

Figure 3.8 Setting a bottle trap. Bottle traps tend to work best set next to edge features such as a clump of vegetation. If the position isn't quite right, for example if the air bubble isn't working or the trap mouth is out of position, it's best to start again: remove the trap and reset it so that the anchor cane is lodged firmly in the substrate. Bottle traps anchored with a cane cannot be used in ponds lined with concrete (it just won't work) or plastic/butyl liners (they might puncture the lining). Instead, the traps can be securely anchored to a weight (David Sewell).

and in places rich in aquatic life that can fill up the trap and quickly deplete the oxygen from the water held inside it. Add an air hole to your trap (preferable) or, if there's likely to be a rise in water level (through floods or rain), contain an air bubble within the trap (see Fig. 3.7). The air hole method is much better (in my opinion), as aquatic reptiles and mammals will quickly use up the oxygen contained in a small air bubble and will therefore expire more easily. If you know that water shrews (*Neomys fodiens*), or similar aquatic mammals, are present at your survey site, you may wish to enlarge your air hole to 2 cm or so to allow mammals that are caught to escape.

Steps must also be taken to ensure trapped animals don't die simply from being trapped for too long. Even if there's access to air, the temperature in bottle traps will rise rapidly in sunny or hot weather, causing thermal stress, and hungry trap occupants may also make a meal of each other! So do make sure you can find your traps when coming to check them. Use a long cane marked

Figure 3.9 Bottle trap line. Canes flagged with red tape have been used to set the bottle traps 2 m apart from each other and to hold each trap at an angle with an air pocket/hole above the pond's surface (Brett Lewis).

with coloured tape to secure the trap and/or attach a floating cork on a length of string as a marker (Figs 3.9 and 3.10). Record your trap locations on a sketch as suggested above (Fig. 3.1) and, most importantly, *count all traps out and back in again.* Worse than the possibility of finding dead animals in your traps is the certain knowledge that there *will* be deaths because a trap is lost somewhere in the pond and is continuing to capture animals because you can't retrieve it.

Bottle trap placement in the water body can be guided by the rule of two (Box 3.1), providing good opportunities both for amphibians to find your traps and for you to compare and contrast your results.

Mesh minnow and crayfish traps work on the same principle as bottle traps but are often double-ended. As with bottle traps, they are usually set and anchored in shallow water (Fig. 3.11) to allow trapped animals access to air via the top mesh at the surface. The trap openings must be under the water. They can be purchased commercially (see on-line auction sites for bargains and suppliers in Section 5.5) and may produce more captures than bottle traps (Madden and Jehle, 2013). A variety of designs are available for crayfish and other fishing applications and are priced starting from just a few pounds (under US $5) each

Figure 3.10 Retrieving a bottle trap. This particular trap was secured with an angled cane marked with tape and also had a cork float attached. At this site, substantial increases in water level are known to occur overnight (with heavy rain) so double precautions were taken to ensure retrieval of the trap. Several newts have been caught (David Sewell).

for collapsible mesh versions. Sturdier, plastic versions cost proportionally more but are likely to last longer in the field. Crayfish-type funnel traps are used commonly in North America and continental Europe, and are gaining in popularity in Britain and elsewhere.

Similar cautions to bottle trapping (including the possibility of starvation or predation of the occupants) apply to the use of mesh funnel traps, so be wary of how long any trap is left in the pond.

Researchers to whom the efficiency of traps is extremely important are always trying to improve their traps' capture rates. Baker (2013), for example, showed that mesh traps baited with pieces of steak captured more great crested newts but not smooth newts than un-baited traps of the same design. Other surveyors found variable results using small light sources (e.g. glow sticks) in the traps: sometimes these appeared to improve captures and sometimes not (see Kröpfli *et al.*, 2010, and Beckmann and Göcking, 2012, for further details). Large mesh traps baited with light have even been trialled for improving captures of large, adult anurans such as American bullfrogs (*Lithobates catesbianus*), with some success (Snow and Witmer, 2011). If your survey project contains enough flexibility, and you have the resources and time to experiment, it might be worthwhile trying

Figure 3.11 Mesh funnel (crayfish-type) trap in position on a newt survey. This trap is staked in position with its upper surface above the water to allow trapped animals constant access to the air. Red-coloured wire frames at each end aid relocation of the trap (Jim Labisko).

food (olfactory) or light baits to see if these work in your particular situation (and many mesh traps come already fitted with a small compartment to contain bait). Un-baited aquatic funnel traps will, however, still catch amphibians.

Ortmann-type traps are also similar to basic funnel (bottle) traps but have usually two or more funnel apertures, either made from bottles or of larger diameter (Fig. 3.12). They can be constructed from plastic bucket-type containers with tight-fitting lids. The funnels are set into the lid and/or sides and the trap is secured at the water's surface (to leave access to the air) with floats or string, depending on the circumstances. As with mesh funnel traps, the multiple apertures can result in more captures, and the large size of the trap holds many captives successfully so long as access to the air is ensured. Ortmann-type traps can be successful for a variety of amphibians, including larger aquatic species (as well as tadpoles and newts) if the funnels are wide enough. For more details and a comparison with other traps see Drechsler *et al.* (2010).

There are still more designs for aquatic traps available, and designs are still being developed and tested. Box traps, such as the Dewsbury trap (Dewsbury, 2011), are similar to an Ortmann trap in that they can have several apertures, though based around a plastic box that is weighted to rest on the pond's substrate. The Dewsbury trap varies, however, in having a large plastic bag attached to the

Figure 3.12 Assembling and setting Ortmann-type traps. A large-diameter funnel affixed to a trap lid can be seen to the lower right in (a). On this occasion, the traps are being used to survey a pond for invasive African clawed frogs (*Xenopus laevis*) in Wales; I'm glad to say none were found on this occasion (a series of very cold UK winters may have finished them off!). Ortmann traps can be secured leaving a surface air pocket, as in (b), or used with floats (e.g. made from empty plastic drinks bottles) to ensure that trapped animals have access to oxygen. Precautions are the same as for other types of funnel trap (Peter Hill).

top of the box that stretches to the water's surface and is kept open to the air with a drinking straw positioned through a float (see Dewsbury, 2011 for design diagrams). In theory (and in testing) this means trapped amphibians can be left in the trap longer without danger of suffocation as they always have access to air, plus the traps can be positioned on the pond's bottom in much deeper water than conventional bottle traps. There is certainly potential here to help ensure the welfare of trapped animals but the resulting apparatus is large and obvious, and may be less suitable for use in situations where interference from passers-by is a possibility. The traps may be considered 'garbage' and removed, or interfered with 'just because'. I tend to use aquatic funnel traps mainly because I'm demonstrating their use to students or conservation practitioners and I haven't yet had a chance to try a Dewsbury trap, because I usually run courses in very public locations: but I'd certainly try the method if the situation was right! Note that, in some jurisdictions (including in the UK), box trapping requires additional licensing as the technique is deemed to be different from conventional funnel trapping.

3.1.5 Dip-netting

Dip-netting is carried out using a net with a 2–3 mm mesh size, a strong handle and robust frame (Fig. 3.13). As with other techniques, dip-netting can

Figure 3.13 Dip-netting in action (David Sewell).

be standardized to provide comparative data (Box 3.2). Nets should be swept through 2-m sections of the pond in the direction you are surveying, again 'chasing' any wary amphibians ahead of you. The technique is most effective when the net is gently agitated through aquatic vegetation and under the pond's edge: there's little point in netting in open water unless you can see tadpoles or adult amphibians you want to identify. Use the rule of two as a guide, and spend, for example, 5 min netting each 2-m section, and record the proportion of the pond's perimeter you've been able to survey.

Dip-netting can be a useful technique because it is normally carried out from the bank and in daylight. It also enables close examination and accurate identification of any animals caught. You will probably want to use a plastic tray with a little water in it to check your captured individuals. Over-enthusiastic dip-netting can, however, cause excessive disturbance and also destroy areas of aquatic vegetation that provide shelter and egg-laying sites for amphibians. For these reasons, it is best to (a) leave it until *after* visual searching, torching and trapping and/or (b) just use it for spot sampling particularly promising areas of the pond to see if there are any species you've missed. If dip-netting is a planned part of your survey it can still be carried out systematically but *always consider your impact on the pond and its inhabitants.*

A further consideration of dip-netting is that the mesh, frame and handle of the net should be disinfected if being used in different ponds more than 1 km apart. Good dip-nets can cost more than £50 (US $80) each and your resources may or may not stretch to several for use on the same day! Refer to Section 2.6 for biosecurity advice.

BOX 3.2 INFORMING CONSERVATION

One of the best uses for robust amphibian survey data is to inform conservation actions. There is still much we need to know to ensure amphibian conservation efforts are effective, and your surveys can contribute to that process.

Beja and Alcazar (2003) surveyed 57 temporary ponds in south-west Portugal that faced a variety of agricultural pressures and were threatened by invasive exotic predators. They used standardized dip-net sampling to detect different species of amphibian tadpoles and adult newts, and compared their findings to pond-specific variables including hydroperiod, potential agricultural impacts and water chemistry.

There were more amphibian species present in ponds that held water for longer, but some amphibians only bred in very ephemeral ponds. Agricultural activities had particularly negative impacts on some species and exotic predators were more likely to be found in ponds that had been deepened to act as irrigation reservoirs.

So a variety of ponds that hold water for different lengths of time are vital to maintaining the range of amphibian species present in this part of Portugal. Farmers like to convert temporary ponds into permanent reservoirs but this study recommends separate reservoirs are created so that exotic predators don't colonize amphibian habitats. Agri-environment subsidies are suggested as a mechanism for compensating farmers who maintain the range of temporary ponds to benefit biodiversity.

Without this study, it would have been easy to assume that the trend to make ponds more permanent would benefit the amphibian community, but of course the reality is more complex and different species do best in different conditions.

3.1.6 Surveying streams

The techniques described above can also be applied to many surveys of amphibians that live or breed in streams and other flowing water, but it's probably more important to research which species you might find in those habitats so you can plan your techniques. Those amphibians that have evolved to use streams are adapted to that environment and employ a variety of methods to exploit it, such as hiding under (or in the lee of) submerged objects or living only in less turbulent areas of water. Visual searches, working your way *upstream*, should include the turning over of rocks and submerged debris to find sheltering adults and tadpoles, and any traps must be set carefully (so as not to be washed away in the flow) in places where they're more likely to work, such as quieter stretches of water or stream pools and side channels.

The data you record from surveys of flowing water have, of course, to be collected in a linear way, so the total length of stream or stream shore surveyed will be noted instead of proportion of pond perimeter. Record the grid references of the start *and* end points of your survey areas for future reference. You can also time-constrain your survey effort to compare different streams or parts of the same stream (e.g. 10 min netting per 5-m length of stream). You should consider whether what you're planning to do is *safe* (assess the risk) and whether it will therefore require more people, time and other resources to achieve your survey goals safely. Stream banks can be difficult to access and the flow rate of the water is much faster in narrow or steep sections. Chest waders and a wading pole can help prevent a soaking, or worse. Again consider the impact of your survey activities, especially if, for example, you want to do an extensive search of streambed rocks and debris. These should be moved carefully and replaced in much the same positions you found them. Holding a suitable net downstream of any rocks you're moving helps prevent hidden tadpoles, salamanders, etc., being washed away before they're identified!

3.1.7 Other techniques

Still more techniques can be employed in aquatic amphibian surveys. Fyke nets, for example, are a type of mesh funnel trap that has a protruding leader panel that guides amphibians into the trap's mouth, increasing its capture area. They can apparently be very effective and efficient but I haven't yet had the opportunity to try them myself: see published studies (e.g. Louette *et al.*, 2013) for more information.

Seine netting is yet another method for use in studies where highly efficient capturing is needed, for example if investigating the relative abundance of tadpoles in a water body. A long, deep net stretched between two end poles is held (by two wading people) at one end of the water body and moved along it while other personnel drive amphibians (and fish, etc.) towards it. The people holding the poles eventually come together, encircling the catch and forming the net into a hammock shape from which animals can be extracted, counted and measured, etc. This requires good personnel resources and some practice! Further information on seine netting is given by Chandler Schmutzer *et al.* (2008) and Preston *et al.* (2012).

A supplementary technique than can be useful for detecting newt eggs in ponds that are perhaps murky, have little aquatic vegetation or are otherwise difficult to survey for some reason, is the use of egg mops (see Fig. 3.14). These are made from strips of black plastic (bin liners) and tied to a cane that is anchored in the substrate to aid relocation. Many newts will readily use such mops for egg-laying, especially if a lack of underwater leaves means the alternative is pebbles or the pond's substrate. Remember, though, that the number of eggs you find *doesn't* equate to newt abundance and, if you decide to try egg mops, you'll have to factor in leaving the mop in the pond to allow the eggs to hatch, and then to retrieve it once newt breeding has finished for the season.

New technologies are providing amphibian surveyors additional and innovative ways of detecting amphibians in water bodies. Environmental DNA (eDNA) techniques, for example, have been trialled in the UK for their ability to detect the presence of great crested newts in ponds and compared with the success of conventional survey methods. Environmental DNA is simply the genetic residue remaining from the presence of target species in the pond. In the UK trial, the eDNA samples were found to be significantly more effective at detecting presence than conventional survey methods and, despite the sensitivity needed to test for eDNA, there were no false positive results (see Biggs *et al.*, 2014). Sampling for eDNA in a pond involves collecting a water sample following a precise protocol, adding ethanol and sodium acetate to preserve any tiny amounts of eDNA present, and then analysis of the sample to test for known components of the DNA of the target species, in this case great crested newts (the procedure is described fully by Biggs *et al.*, 2014).

Figure 3.14 An artificial egg mop ready for setting. The strips of plastic replicate the leaves or fronds of aquatic plants and are readily used for egg-laying by, for example, newts. Eggs deposited on these mops should be allowed to hatch, after which the plastic needs to be removed from the pond. In lined ponds where a cane might cause damage, you can use a weight to anchor the egg mop and a string attached to a cork float to aid retrieval (John W. Wilkinson).

Because great crested newts are widespread in the UK, but at the same time enjoy European-level protection, large amounts of money can be spent on surveys and obtaining population estimates when the species' presence holds up land development plans: so the UK government has high hopes for eDNA sampling as a shortcut alternative to full surveys. Though the technique seems to be very good at revealing species' presence, it cannot (at the time of writing) be used to infer population sizes. I personally hope that eDNA sampling will become part of the repertoire of amphibian surveyors and be used to identify and target ponds where detailed population surveys need to be carried out. In this way (hopefully), important amphibian populations can be monitored and protected, and the survey skills key to understanding biodiversity conservation will not be lost!

Finally, even if you've planned a survey based on only aquatic survey methods, don't ignore rotting logs or other terrestrial refugia near the pond. In spring 2013, a colleague and I took a group of students to survey a pond that looked as if it should provide good data, and we thought we knew which species ought to be found there. Unfortunately there had been a few days of cold weather and we were unable to detect any amphibians at all! So, while my colleague demonstrated the setting of bottle traps, I searched under a few fallen branches nearby and was able to find juvenile newts of two species. This prompted us to return to the pond later and we eventually found adult female newts laying eggs, as well as several toads. I've also added to a pond's species list by finding amphibian remains left uneaten by an unknown predator. So, you should always use all your observational skills during surveys: any evidence of amphibians can inspire you to search maybe just that little bit more and locate the species you had hoped to see!

BOX 3.3 INCREASING YOUR CHANCES

Increasing your chances of detecting all species – and having high confidence that a species is *absent* (as opposed to simply undetected) – in your survey pond are *both* important for accurate results, and so improve confidence in your survey overall.

Sewell *et al.* (2010) carried out a series of amphibian surveys in order to test the methods of the UK NARRS. They trialled four aquatic survey methods (visual searching, torching, funnel trapping and dip-netting) in two separate areas over two years and used occupancy modelling to examine how successful the methods were at detecting the presence of five widespread amphibian species.

Their results revealed differences in detection success between species, areas and years but, most interestingly, showed that *using all four survey methods* could reduce the number of surveys needed to be confident a species was present or absent. The method that made the most difference to detection success was funnel trapping, especially for newts, though the authors note the extra precautions required to carry this out (see Section 3.1.4).

In other words, then, the more survey methods you can use the more chance you'll have of detecting all the species present in a given water body. Occupancy modelling can be complex to carry out and interpret, but this very useful study resulted in a recommendation that NARRS amphibian surveys include all four methods and at least four visits whenever possible.

3.2 Amphibian surveys in terrestrial habitats

3.2.1 Preliminary visits for terrestrial surveys

Although the terrestrial surveys you carry out may be some distance from water and slippery banks (but not necessarily), it's still best practice to get to know your survey area well before you start. Sketch the locations of any orientation features or hazards in relation to the transects or plots you intend to survey, as well as any trap locations (if used), and then use your sketch to ensure safety and for reference and reporting. These days, of course, the more sophisticated GPS units and smartphone mapping apps allow you to add and store a good deal of information, including tracks and features. If you're familiar and happy with using this technology: why not!? But if you're relying on electronic devices, as usual don't forget the obligatory spare batteries and make sure you're able to download any maps you create to use in your reports.

Data about the terrestrial habitats in which you are surveying should also be recorded before the important business of looking for amphibians starts. To work out what habitat and survey condition information you'll need to record, again think about formulating the questions that your survey will answer. Some suggestions are given below and in Section 3.3. This can help you during the surveys, to compare the characteristics of different sites, and afterwards, to refresh your imperfect memory!

3.2.2 Surveying terrestrial plots

Without the focus of an aquatic breeding site, the terrestrial areas you survey must be defined in some way. This could be a pre-determined 'site' (or sites) in which you're interested (say a particular field or nature reserve), different habitat types within a site or similar habitats at different sites. In itself, defining a survey site can be problematic because amphibians (and other organisms) don't respect fences, land ownership boundaries or site designations. These concerns are not, of course, unique to amphibian surveys.

You can attempt to make your terrestrial survey results as robust and comparable as possible by taking representative, random samples throughout the areas you're surveying. One way of approaching this is to use terrestrial plots. These are plots of a consistent, defined size (say 5 m × 5 m) that are measured and marked out at least 24 hours in advance of surveys (so that there has been time for any amphibians present to settle before you start searching them). Mark out plots using sticks and string, dabs of paint or tape, or you can even fence the plot to ensure no animals can leave it while you're searching. The types of fencing used in drift fences (e.g. Fig. 3.21) are suitable for this.

Terrestrial plots are most useful for examining habitat use by amphibians outside their aquatic breeding season, or for totally terrestrial amphibians that have no need of water bodies for reproduction, such as plethodontid salamanders and *Eleutherodactylus* frogs (see Box 3.4 on page 63).

Plots of this type in forest habitats are often called *leaf-litter plots* because searching them involves combing carefully through the layers of leaf litter on the forest floor (see Fig. 3.15). More generally, though, you should carefully examine the litter and other natural refugia (logs, rocks, etc.) within the plot, starting at the edge and moving towards the centre of the plot. Any amphibians encountered should be kept in moistened cloth bags – or similar holding containers such as plastic boxes containing some moss or leaf litter – until the whole plot has been searched. This reduces the chance of spooked animals escaping and prevents you from counting the same animals more than once (which is important to avoid), as well as allowing species identification and any other recording to be carried out at a more relaxed pace. As with any survey activity, you'll need to consider H&S, wearing suitable gloves, strong clothing and boots to reduce cuts and other injuries, especially in locations where venomous snakes or scorpions, etc., could be encountered. Ideally, you should research any venomous or stinging animals that may be present in the area and perhaps carry a laminated, quick reference picture guide to these. Avoid any creatures of which you are in any doubt! Small hand rakes or sticks can also be useful in the search.

Figure 3.15 Surveying a large terrestrial (leaf litter) plot in the Seychelles (Jim Labisko).

The size and number of terrestrial plots you survey may depend upon your available resources: larger plots require more time and usually two or more people to search per plot. Consider that 32 plots of 5 × 5 m spread throughout a habitat will give you a larger sample size (in terms of the number of individual samples) and each plot will be easier to survey than, for example, eight plots of 10 × 10 m, though the terrestrial area actually surveyed is the same (800 square metres) for both survey designs. In order to randomize your sample design, you may want to create a numbered grid on a map of the habitats in your survey area and then use a random number generator to choose which plots will be surveyed. Alternatively, you could survey plots at random distances along a transect (see Section 3.2.4), recording the differences in habitat around each plot. As usual, your sampling approach should be refined depending on the survey question you're attempting to answer.

3.2.3 Using artificial refugia

Though some small terrestrial anurans and salamanders can occur at high densities in their preferred habitats, amphibians that breed in water can sometimes be difficult to detect outside the breeding season and away from their breeding ponds or streams. So the plot survey approach described above can be augmented by placing artificial refugia to increase detection rates. These may be made from materials similar to those used commonly in reptile surveys, such as corrugated steel sheets, roofing felt or carpet tiles. Artificial refugia are sometimes called artificial cover objects (ACOs). Whereas reptiles use such refugia to aid thermoregulation, amphibians are more likely to exploit them as a moist shelter (Fig. 3.16). The placement and type of refugia are therefore important. Only use materials likely to get very warm (such as corrugated steel) when in cool, shady locations where amphibians will find and use them. Roofing felt or carpet tiles are less likely to get too hot; they are effective for amphibians and both these materials retain moisture, as well as being lighter to carry than metal sheets! Placement of refugia through a habitat can be on a grid system or array (space each refuge say 5 m away from one-another), randomly located throughout a habitat in the same way as terrestrial plots, or along transects of the type described below (Section 3.2.4).

It is generally accepted that refugia left to 'bed in' (any vegetation underneath becomes flattened and the refuge establishes its own microclimate), and for animals to find them, for a few weeks or more, will yield better survey results and their effectiveness will probably improve over time. Refugia left in place long-term, however, may become home to nests of ants, other insects and/or small mammals and may require relocation to a place nearby if your study takes place over a relatively long period. Once relocated, they will again benefit from bedding in.

Figure 3.16 Juvenile toads (*Bufo spinosus*) found sheltering under an established artificial refuge in Jersey, Channel Islands (in this case a sheet of tin sheltered beneath a bramble patch). The surrounding habitat is heathland, often very hot and dry in summer, where amphibians may otherwise be very difficult to detect outside their breeding season (John W. Wilkinson).

Artificial refugia can also be used in surveys for arboreal species (e.g. Boughton *et al.*, 2000; Fogarty and Vilella, 2001; Johnson *et al.*, 2007; and Box 3.4 on page 63). Use ca. 5-cm diameter sections of plastic (PVC) piping cut to 15 cm or longer lengths, or similarly sized lengths of hollow bamboo. These can be wired to branches at different heights, in different locations and different habitats for a variety of comparative studies. I once visited an Amazonian reserve where other researchers were carrying out a study of some sort (not on amphibians) using pipes of about 1-m length stuck into the ground at intervals of several metres apart: each pipe had its own resident population of *Scinax* spp. and other treefrogs. It was a very interesting way for me to find some of the amphibians present in the area!

3.2.4 Visual encounter surveys

Visual encounter surveys (VES) in terrestrial habitats, also often referred to as visual transects, depending on their exact nature, are essentially similar to visual surveys of aquatic habitats, though usually without the focus of an aquatic breeding site. As with any other technique, comparability – and therefore equal effort – is key to answering your survey questions. You can use VES to investigate differences in:

* species composition
* species richness
* relative abundance
* habitat use
* activity periods of different species

between transects in different habitats or *along* transects where there is a gradient of change (e.g. from wetter to drier conditions or with altitude). In any case, each transect must have habitat data associated with different points along it so that the data can be related to any differences observed in your survey results.

As an example, you might want to compare which species of amphibians live in stands of native mixed woodland and stands of planted timber trees. You could compare the two habitats by surveying a 300-m line transect in each habitat (a *distance-constrained survey*) or by searching for 2 hours in each habitat type (a *time-constrained survey*) but there will be lots of unpredictable variations between the different forest stands. How easy is it to walk down each transect? How far can you see to either side in each habitat? Are all the species you might encounter equally detectable in both forest types? Another option is to compare the amphibians detected on different transects using *time- and distance-constrained* surveys: decide on a transect length and over what time period you will survey it, and stick to that combination for all the surveys you're comparing.

Where your survey study has the time available, several (or many) transects with identical constraints can be set up to increase the amount of data generated by your survey and improve confidence in the results. Record environmental and habitat (vegetation, insolation, moisture, altitude, etc.) information and any changes in habitat in the same way along each transect: perhaps at 5-m intervals (see also Section 3.3).

As well as comparing different habitats or gradients, you could of course carry out repeat surveys of the same transect at different times of day or night, or at different times of year. The ecology of different amphibian species may result in different species being active or visible at particular times over a 24-hour period and, of course, at different times of year. So if you're doing several transects on the same day, don't always do the same one first or you could bias your results! Rotate the order in which you start each transect or use a random number generator to decide the order in which the transects are surveyed.

Night transects (usually the most productive as most amphibians are more active at night) will of course require a torch or torches. One of the many excellent

LED head torches now available is usually adequate for terrestrial VES surveys at night – so long as the batteries are fresh – as opposed to the more cumbersome units needed to survey ponds (see Section 3.1.3 and suppliers in Section 5.5). I'd always still carry spare batteries *and* at least one other spare torch though!

It can in fact be quite difficult to survey over the same distance, within the same set time period, in contrasting habitats with variable terrain, but try to avoid, if possible, using only tracks, paths or habitat edges for your surveys. Although these are always interesting elements of habitats and might be easier to travel along, your results might only indicate which species occur along those linear features (or which are most easily detected there) if they are used exclusively. Conduct a preliminary survey during good daylight (and when you are *not* under pressure from time constraints) to consider these issues. If you think avoiding bias will be difficult, use a good map to identify random but suitable starting points in each habitat type, then set up your transect along a random compass direction from each starting point. At the same time, however, be practical (see Fig. 3.17). Choose a replacement transect route if there are unacceptable hazards or if the route is too difficult or time-consuming to survey (because dealing with

Figure 3.17 Your survey transects must be practical and achievable. A 500-m transect through this steeply undulating Pacific island rainforest could take far too long to walk, let alone survey adequately. Nevertheless, a 90° change in compass direction would allow for a flatter transect through the same habitat from the same starting point. It should also not be excessively difficult just to reach the starting point of the transect. You should always try to remove biases from your comparative surveys: but never set yourself up to fail! (John W. Wilkinson).

these things introduces biases anyway). Mark out transects with high visibility markers (reflective tape or cloth tied to sticks or twigs) and/or be scrupulous about taking GPS readings and recording trackways (and have spare batteries) so that you don't get lost. This will also help ensure that repeat surveys of the transect really are along the same route. Don't forget to take a GPS reading of your base location, or the location of your vehicle, too! As with aquatic surveys, for both safety and practical reasons, such surveys should be carried out by at least two people working together, with responsibility for different elements of the survey (spotting, recording, taking GPS readings, etc.). Be clear, however, about who has which responsibilities.

Actually surveying your VES transect consists of walking along at a suitable pace, examining the open ground and looking in or under any hiding places, as well as on branches and leaves in locations where arboreal species could be encountered. Your planning will, as usual, be helped by a good identification guide obtained prior to departing for the field. What species are you *likely* to encounter? If the appropriate field guide is too big to carry with you, prepare a laminated species list (with pictures if necessary) to jog your memory while surveying.

Exactly what you record about the amphibians you encounter will depend on what your survey question is, but species, age (adult/juvenile), sex, substrate (ground, leaves, etc.), distance along transect and time seen are all useful data. Design your survey form to fit your needs (see Section 5.1). I would strongly recommend the use of a good-quality digital camera (most models now tag each photo with the date and time) to reduce the information that has to be recorded manually while actually surveying. In very wet areas or during rainy seasons, a *waterproof* digital camera is best. If there are several similar species present in the area, be sure to photograph their diagnostic features. Even if (when absolutely necessary) you have to take specimens back to base for definite identification, captured animals should be released at the place they were caught, and in similar conditions (e.g. don't abandon shy nocturnal treefrogs in the midday sun and heat).

Conducting VES constrained by both time and/or distance should help with comparability later on, but the search effort needed in different localities can still be highly variable. What if one habitat has lots of large fallen branches under which you want to look as you walk the transect: will you have enough time to get to the end within the pre-determined time period? Though a VES transect has a finite length, I'd suggest you also imagine it as an invisible tube perhaps 4-m wide (2 m either side of your transect line) and about 2-m high and that you aim to detect animals located on the ground or in vegetation only within this imaginary tube (Fig. 3.18). The size of the imaginary tube can be varied depending on the terrain, habitat and species with which you're dealing but it should be consistent between comparative transects. It will require some practice not to deviate from this (and if you do get the time to practise before 'properly' starting surveys, you

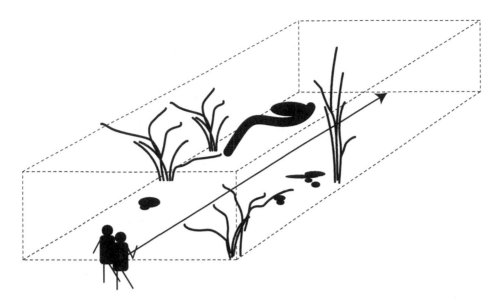

Figure 3.18 Diagrammatic representation of the start of a VES transect within an imaginary rectangular tube about 4 m wide and 2 m high. Only areas of ground, hiding places and vegetation within the tube are examined, and amphibians seen outside of it are excluded from comparative survey results (though, of course, 'extra' species records can also be noted down). The black arrow represents the path of the transect line. Even a robust and random transect doesn't, of course, *need* to follow a precisely straight line but, if your transect has to change direction at one or more points along its length, make sure these changes (and where they happen) are recorded and well-marked in the field (John W. Wilkinson).

really should) but thinking through your survey approach in this way will help ensure you achieve your objectives and answer the survey questions. After all, all the planning in the world won't help comparability if you get so absorbed in searching a particular tree that you never reach the end of your transect line! Again remember your impacts on the habitat and avoid the total destruction of natural hiding places (e.g. rotting logs, loose tree bark) where possible. Always replace things much as you found them, especially if your transect will be re-surveyed later on.

3.2.5 Pitfall traps and drift fences

Pitfall traps, consisting usually of bucket-sized containers sunk into the ground so that the rim is flush with the ground's surface, are a passive way of sampling terrestrial areas. They can be very effective (Fig. 3.19), though of course they work best for species that live only at ground level. Any amphibians with good climbing ability can usually find their way out! Ensure there are no gaps between the outer walls of the trap and the substrate, as small amphibians in particular will fall or squeeze down and may be undetected, or even harmed.

Figure 3.19 A pitfall trap haul of toads (*Rhinella crucifer*) from a survey in Brazilian Atlantic forest (Moacir Tinoco).

Pitfall traps can be deployed in a grid or array (as with refugia) or within plots or along transects (as described above) at consistent densities or distances from one-another.

This type of trap is, however, more effective when used in conjunction with drift fencing (see Figs 3.20–3.22). Amphibians travelling along the ground encounter the fence and are directed along it until they drop into your carefully located pitfall trap. A variety of designs of drift fences can be used, either for stand-alone sampling of amphibians in different terrestrial habitats or in association with features such as breeding sites. The drift fence barrier can be constructed from stiff plastic sheeting or even from chicken wire mesh sleeved with plastic bin liners, held in place with wooden stakes. Several centimetres should be sunk into the ground to prevent amphibians burrowing underneath. Proprietary drift fencing is also now available (see suppliers in Section 5.5 and Fig. 3.23) though will inevitably be more costly than home-made designs. Pitfall traps should be spaced at convenient (and consistent) intervals for checking, say 2 m or 5 m apart, depending on the length of your fence/s. You will, of course, also need to have tools to construct the fence (e.g. a staple gun or screws and washers to affix the plastic to the stakes, and a mallet to knock the stakes in) and bury the traps (a trowel or spade), so you'll need to consider the need to carry these around when setting up your trap line. In exposed areas, more fixings (and

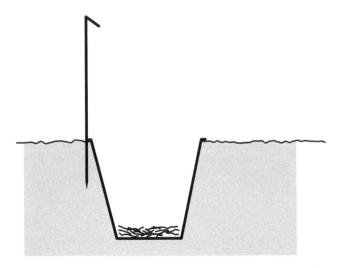

Figure 3.20 Cross-sectional diagram of a pitfall trap in use with a drift fence. Note that the trap is flush with both the ground and fence. If surveying in an area where fossorial caecilians could be encountered, the top of the pitfall trap will need to be perhaps 30 cm *below* the ground surface (i.e. leave a 'wall' of the substrate around the trap's top) to allow them a chance to wriggle into the trap*. Some proprietary fences come with an angled lip (as shown) to help deter any animals trying to climb over. For details see text (John W. Wilkinson).

* Probably the best way to survey specifically for fossorial caecilians is to dig carefully through the leaf litter and substrate. Such surveys are somewhat specialized and unusual, so see published papers for ideas and guidance.

more robust ones) will be needed to keep plastic fencing affixed to stakes: windy conditions will pull the fence away from mere staples. As with many other survey tasks, building a drift fence is often not a job for just one person!

The locations of traps should be well recorded and numbered, and each trap should have puncture holes at the bottom so that rainwater drains away (and you could add a polystyrene float to help prevent animals drowning in the event of heavy rain). In locations where traps are exposed to full sun or other extreme weather, it may be desirable to use them in conjunction with refugia (place each refuge over a trap, raised slightly on stones or sticks to shelter it while allowing amphibians to crawl underneath). Some moss or leaves in the bottom of the trap will shelter and help calm any captives waiting for collection but ensure there's not too much material in there that will help your captives find their way out. On the other hand, an 'escape stick' or mammal ladder (available from fence and trap suppliers) can be added to allow the egress of small mammals (which might otherwise die, or even eat your captives, before they can be released).

Pitfall traps need very careful checking! *Look* before digging around with your hands, especially in low light conditions, to check for venomous snakes, biting invertebrates, shrews, etc., and be prepared to have to extract and deal with

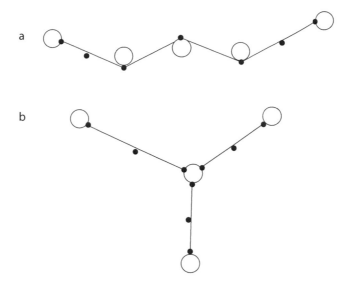

Figure 3.21 Example arrangements of drift fences with pitfall traps. Large open circles show pitfall trap positions and small black circles represent stakes (more or fewer of which might be needed depending on the stiffness of the fencing material). Fence bases must be buried, and traps must be flush with the ground *and* the fence to be effective. (a) Linear pitfall line – can be set out perfectly straight but tends to work slightly better (in my experience) when 'kinked' slightly, as shown, so that travelling amphibians are funnelled slightly towards the traps. Fence section length and the number of traps can be varied according to needs and resources but should remain consistent in comparative studies. Linear trap lines can also be used to encircle breeding ponds for sampling all ground-dwelling amphibians arriving to breed. In such a survey, it will usually be necessary to also place traps on the *inside* of the fence, as some species like to move in and out of ponds over the course of the breeding season. Pairing pitfall traps in such an arrangement (for every trap on the outside of the fence there's also one on the inside) can reveal a lot about amphibian movements; the direction from which they arrive to breed and where they leave the pond – but the resources needed for this kind of study (and to check traps responsibly) can be considerable. (b) An extension of a linear trap line is a star-shaped array, which is more appropriate when there's less of an idea about the direction in which amphibians may be travelling, for example through a forest clearing. Larger arrays with more 'arms' can be used if space and resources permit (John W. Wilkinson).

these before you examine your intended captures. Be sure any such creatures are well away from the trap line before you record the amphibians! As with other amphibian survey techniques, pitfall traps will often work better left overnight but don't forget that many amphibian species are diurnal: it is good practice to check traps at dusk and dawn (at least) and to modify your trap-checking schedule to account for local species differences. If traps are operating during daytime, more frequent checking will help ensure that captured amphibians remain healthy during hot or sunny weather, or under any conditions of extreme change.

If pitfall trapping is an important part of your survey but (for whatever reason) you will need to leave the traps unchecked for a day or more, choose to use buckets that come with close-fitting lids. Traps can also be closed down

Figure 3.22 Drift fence and pitfall trap array being used to sample herpetofauna in a dune system in Atlantic coastal Brazil. Digging the holes for traps and anchoring stakes will, of course, take relatively longer in forested or rocky habitats! (Moacir Tinoco).

Figure 3.23 Proprietary drift fencing staked in an extensive grid pattern. Here, drift fences are being used in conjunction with both refugia and pitfall traps to clear an area of amphibians prior to a development, but a similar arrangement (usually on a smaller scale) could be used to capture amphibians for a survey project if the resources were available. Drift fencing can also be used more modestly to bound individual terrestrial plots but are prone to vandalism and/or damage from people, livestock and, for example, falling tree branches. They should not, therefore, be used in locations where these factors will either increase the chance of harm to your captured individuals or affect the efficacy of your trap line (S.A. Graham).

temporarily by filling with sticks, moss, leaves and other handy debris, so that trapped animals can quickly climb out. Responsible surveying means taking every step to ensure you don't cause needless suffering or death, or any other adverse impacts on the species and habitats in which you're interested.

BOX 3.4 EVALUATING TERRESTRIAL TECHNIQUES

The survey techniques you use in your own surveys can always be informed by previous studies. Refer to others' surveys to help decide how best to answer your survey questions.

Fogarty and Vilella (2001) carried out mark–recapture population estimates, calling male line transects and point counts of *Eleutherodactylus* frogs in native forest and managed plantations in the Cordillera Central of Puerto Rico. Frogs were trapped for mark–recapture using artificial arboreal refugia (bamboo) arrayed in a 33-m square grid in both types of forest, and marked with fluorescent pigments. Ten 50 m line transects overlaid mark–recapture grids in both types of forest and each transect was monitored on up to 30 nights. Calling males within 5 m of the transect, and their distance along it, were recorded. Additional estimates of total numbers of calling males were obtained from 2-min point counts at the centre of each mark–recapture grid. Analyses of the data obtained were carried out with specialist software.

The density of *E. coqui* males was greater in native forest, but more males called in both forest types when relative humidity exceeded 80%. Only one species was detected using arboreal refugia, whereas transects and point counts were both able to detect three species. The authors also found that (brief) point counts of calling males did relate to density estimates of calling males obtained through (relatively very time-consuming) mark–recapture and line transects. They recommended further studies on the relationship between calling surveys and population estimates.

This is a fairly complex study using multiple terrestrial methods; it begins to reveal the relationships between different survey methods requiring very different effort. The point I hope to reinforce with this example is *don't be afraid to stand on the shoulders of others!* Research the survey methods you could use to provide appropriate data with the effort you are able to expend on it (see also Section 3.2.6 for more information on calling surveys). This also allows you to compare your results with those of other studies: always useful in helping funders and decision-makers interpret your findings. You can (and probably should if, like me, you are not a statistical expert) also research analysis methods published by others for suggestions as to how you could carry out and present analyses of your own.

3.2.6 Calling surveys of male anurans

This technique is covered under the section on terrestrial methods because it does not require aquatic survey equipment (such as nets). Nevertheless, calling surveys can be conducted at or around breeding sites, as well as in completely terrestrial environments (for species without an aquatic stage and so not associated with water).

If you're familiar with the species likely to be present in your survey area, and can recognize and distinguish their call, noting the presence of any male anurans heard calling can add to your species inventory because it isn't essential to see the animal calling to know it is present. Calling males can be heard and/or recorded sight-unseen when in the forest canopy, calling from a hidden burrow, or simply if you can't actually find them! Recordings of calls can also help identify the presence of different species that look superficially very similar: these can have very different calls from one another. Conversely, if you *don't* know which species might be present, or if there isn't enough information available on the area you're surveying, sound recordings of the species calling can allow you to research the range of species you're hearing *after* the main part of the survey is over. This can be particularly valuable if you're seeing different species of frogs or toads, and hearing calls, but can't relate the calls to any of the animals you've seen (this has happened to me!). Cross-reference your recordings with published or on-line audio resources and sound libraries. With high-quality digital recordings and the use of analysis software, calls can even be teased-out and identified to species using sonograms that enable visual distinction between calls (e.g. Platz, 1989; Kime *et al.*, 2000). Some anurans, of course, have a repertoire of different calls, used to establish a territory, attract females or to deter other over-amorous males, and some calls sound very different under different conditions of temperature and humidity. Be careful not to assume that these calls belong to different species: using as much published information as you can lay your hands on will again be very helpful here.

Recording equipment could be as simple as your smartphone (especially in an emergency) or involve specialized digital equipment and a shot-gun microphone. If anurans that call underwater could be encountered, a hydrophone will be required. Take a range of clear plastic bags and insulation tape to protect your equipment from rain and damp (as well, of course, as spare batteries!). In any case, note the date, time, location, habitat and weather (especially temperature) conditions for each call you record, as well as whether you think other background sounds (including of other animals) are present on your recording. Some insects and (especially) geckos make calls that are frustratingly frog-like!

Audio transects of calling males can be an important part of some surveys (see Heyer *et al.*, 1994 for detailed information) but it's often difficult to determine metrics such as the distance of the calling frog from a transect line, or even how many callers there actually are, without a great deal of effort, practice and experience. To simply add to your species inventory, Lips *et al.* (2001) recommend that 'listening points' are located 100 m or more apart, to avoid recording the same animals twice, and the use of an index of 'number of males calling'. These authors propose 1, 2–5, 5–10 and more than 10 calling males as suitable categories with which to compare relative abundance, and these four divisions should be suitable for most purposes. See also https://www.pwrc.usgs.gov/naamp/ for the protocols of the North American Amphibian Monitoring Program (NAAMP), which uses a roadside transect and emphasizes the importance of training in call recognition. For some species, and especially in the tropics where several species may be calling simultaneously and rapidly, it can be difficult to assess numbers of males calling when there are any more than one!

Automated recording systems can also be used, and set up and left overnight or longer, to capture the calls of the species present. These can be a useful and time-saving tool when used in conjunction with automated environmental data loggers. As well, however, as the need for possibly expensive equipment that has to be water- and weatherproof, and have a suitable power supply, there's the possibility of interference and theft to consider. The setting up of automated recording stations therefore needs to be well-resourced, and each station needs careful siting, or hiding. More information about automated call recording is given by, for example, Bridges and Dorcas (2000) and Acevedo and Villanueva-Rivera (2006).

Note that only the relative abundance of calling males can be determined however much effort is put into an audio survey. Salamanders and some anurans don't call at all (or make very subtle sounds), and females and juveniles of different species will probably be present in any given habitat in different ratios to their calling males! So, though surveys of calling males can be a very useful technique, it will not inform your survey about absolute numbers of amphibians.

BOX 3.5 SOME FINAL SURVEY ADVICE?

I haven't yet mentioned one of the potentially most useful techniques you might use on your surveys: asking people! Whether an e-mail to the nearest university biology department or a chat over a beer with local people, don't be afraid to ask folk if they've seen or know of any amphibians where you're surveying. This may involve pictures in a guidebook and lots of pointing if you don't speak the local language (though efforts to do so are usually appreciated). Either way, you might just find out something that will help you acquire information. This advice comes with a warning, however: some people can be so willing to help that they may have 'seen' species that cannot be present within thousands of miles: so maybe try a picture of an amphibian from another continent, if you can do so subtly, just to gauge how reliable your informant really is …

3.3 What other data should you collect?

In Section 2.4, I discussed the need to record *what, where, when* and *who* (recorded it) to create a valid biological record. These data remain the minimum that should result from any amphibian survey. I've also been emphasizing the value of comparative information: *what makes your amphibians tick* can be thrown into sharp relief by comparison with other surveys, other sites and with any measurable changes over time. Long-term continuous datasets on amphibian populations are rare, but carefully collected data separated by several (or many) years can also be very valuable as long as the comparative information included is quantified consistently, presented transparently and made available to other surveyors (see Section 4.2). The *other* data you collect is really as important as the amphibians you record, even if it's only there for future surveyors to compare their data against! So if your goal is simply to know which species of amphibians are present in a particular nature reserve, you should always assume someone – even 20 or more years down the line – will have the same objective, and that information then *becomes* comparative. What's required to explain any observed differences between two or more sets of amphibian survey results (whether separated by time or location) is *what else varies* between the surveys.

Another important reason for collecting additional data is to demonstrate that you carried out your survey under appropriate conditions (which other surveyors can repeat). For example, a survey of an English woodland pond in December finding no amphibians would (and should) be discarded: it was carried out under inappropriate conditions (at completely the wrong time of year). Yet, sometimes, such surveys are presented (occasionally by unscrupulous developers or even ecological consultants) as evidence of the absence of amphibians, though the same survey in March may have detected three or more species. So *environmental data* can be key to improving the comparability (and robustness) of your survey.

If, however, your survey question is to determine, for example, why one pond supports more amphibian species than another, collecting *habitat* data is also important. How are the habitats present at each site different from one another? You may not know, initially, which species are present or, indeed, what habitat preferences those species have, so you should try to collect habitat data in an impartial manner as far as is practical and possible.

Metrics and methods for collecting both environmental and habitat data are suggested below. Some variables don't fall neatly into either category (e.g. pond depth) and may need to be recorded on each survey visit (because they can change) whereas others will remain constant throughout your survey period and need recording just once. I've tried to divide up the suggested variables as logically as I can in the tables below but you can of course make your own decisions based on your particular survey questions. Most important is to be consistent and pragmatic, with the resources available to you, and to make the effort to record *all* the data *you set out to* record in every survey.

Additionally, Habitat Suitability Indices (HSIs), such as the one developed by Oldham *et al.* (2000) for the great crested newt in Great Britain, can be used as a more rapid way of quantifying comparative variables. The development and use of robust HSIs takes considerable research time but can result in an objective way of assessing habitat and other parameters. More HSIs for other species are being developed at the time of writing and will no doubt prove their utility in due course and with practical trials.

3.3.1 Environmental data

Table 3.1 lists a range of environmental variables you might want to collect during an amphibian survey, and any equipment you'll need to collect them. See also the example survey forms in Section 5.1 for suggestions about the ways these data can be recorded effectively.

Environmental data recorded *prior to* surveying can also be revealing regarding amphibian activity patterns. The most obvious example of this for amphibians is rainfall within the previous day or days, but many factors can have an influence. The average temperature, for example, can influence emergence from hibernation and start of breeding (e.g. Reading, 1998). In practice, most surveyors won't be able to set up a full-blown weather station weeks or months before surveying starts, but all is not lost in this respect as local weather data can sometimes be obtained for prior periods from the nearest meteorological station or be available as geographic information system (GIS)-compatible datasets (see below). Though this approach is inferior to using data actually obtained at your survey site, it can help reveal influencing factors.

Additional information to include that isn't strictly 'environmental' is the start and finish times of your survey. This will help assess the effort put into different surveys and can be used to generate further comparative measures, for example *Surveys at two farmland ponds recorded five frogs per hour spent surveying at Pond A and eight frogs per hour spent surveying at Pond B*. The ponds may be different sizes or surrounded by completely different habitats, but the amphibians present at each (in this case the rate at which they were encountered) can still be usefully compared because the effort spent surveying was recorded. A higher encounter rate, assuming no differences in frog detectability between the two ponds, suggests that frogs are more abundant at *Pond B*. Ensure any possible issues with detectability, such as turbid (murky) water or lots of aquatic plants, are noted on your survey form so that you can at least discuss these in the context of any apparent differences in your report, and so that this information is available to others.

Table 3.1 Suggested environmental variables for amphibian surveys.

Variable	Units or scale	Equipment needed/notes
Air temperature/ ground temperature (for some species, subsoil temperature may also be useful)	°C	Traditional or digital thermometer (with probe). Standardize the height at which air temperature is recorded. For detailed surveys record temperatures at the start and end of the survey session
Water temperature	°C	Traditional or waterproof digital thermometer
Relative humidity	% humidity	Hygrometer (often combined with digital thermometers*)
Water clarity/turbidity	Categorical scale** or nephelometric turbidity units (NTUs)	Turbidity meter (nephelometer) if not using categorical scale
Rain	A categorical scale** is often most practical during survey, otherwise mm of rainfall	None, or rain gauge if recording rainfall over a set period
Wind	Wind speed in km/h or Beaufort scale/simplified scale**	Handheld anemometer if using km/h
Moonlight/sunlight	Lux or categorical scale**	Digital light meter*
Lunar phase (some researchers assert that lunar phase influences activity in some amphibians, others disagree!)	New/first quarter/ full/last quarter moon	None: often shown on calendars or watches, or can be found on the internet
Cloud cover	Oktas (0–8) = how many eighths of sky are obscured by cloud	None or weather station (may be superfluous if measuring light levels)
Barometric pressure (influences calling in some anurans)	Millibars	Handheld barometer*

Variable	Units or scale	Equipment needed/notes
Altitude (at different transect points or if inventorying multiple sites)	Metres above sea level (m.a.s.l.)	Altimeter or GPS unit (won't have changed if you're re-surveying the same pond!)
Water pH	pH scale	pH meter
Other water chemistry parameters (e.g. dissolved oxygen, water hardness). Some species have very specific requirements but this won't be needed for most surveys	Specific to each parameter	A variety of test kits/meters are available (see Section 5.5)
Flow rate (e.g. of streams)	Volume of water per unit time	Current meter (this is almost a habitat parameter but may vary greatly between survey visits, so it's useful to think of flow as a changeable environmental variable)

* Some suppliers (see Section 5.5) now sell combined thermometer/hygrometer/anemometers that also take other measurements. These can save space but are expensive initially and have (in my experience rather often) an annoying habit of one function not working after a while. Individual meter units are probably more cost-effective in the long term if you need precision (though many digital thermometers come with a hygrometer).

** Depending on your resources and available time, simplified categorical scales can sometimes be used. See Section 5.1 for examples of how some environmental variables are recorded in UK NARRS and other surveys.

3.3.2 Habitat data

If you've carried out a preliminary visit to your survey site and made a sketch, you could be well on the way to knowing what habitat data should be recorded! These data do need quantifying, however, to improve comparability. Table 3.2 suggests some habitat variables you may want to collect and examples of how they can be recorded. The list is not exhaustive but should give you some ideas about how you could record habitat parameters during your own surveys. See also the example survey forms in Section 5.1.

Habitat (and environmental) data, including vegetation cover types, linear connectivity and barrier features, etc., can also be associated with your survey results through the use of GIS techniques and (sometimes freely available via the internet) habitat datasets such as national or international land cover data.

This can save time and effort in the field but requires desk processing time (and, if you're like me, the help of someone who knows how to do it efficiently!). Nevertheless, the use of GIS can also make your survey results useful for projects such as predictive mapping of species distributions (see also Section 4.4.3).

Table 3.2 Suggested habitat variables for amphibian surveys.

Variable	Units, etc.	Notes
Pond size, perimeter length, pond depth	m² or m	Most useful in conjunction with a sketch showing pond shape. These parameters can change quickly with drying or rainfall and may need to be recorded on each survey visit
Pond/stream substrate	Type/s and size (if appropriate)	e.g. mud, sand, gravel, rocks <10 cm diameter
Estimation of proportions of different (broad) habitat types at survey site or within a certain distance of pond (e.g. 250 m, 500 m)	Visual estimates or measured areas (%)	e.g. 10% bare ground 35% grassland 40% woodland 5% woody debris etc.
Detailed record of habitat or vegetation cover and heights at survey site or within a certain distance of pond (e.g. 250 m, 500 m)	Measured areas (m²)	e.g. broadleaved woodland 56 m² rocky scree 120 m² mixed grasses 40 m² etc.

or oak woodland >8 m high 230 m² *Molinia* grassland 40–50 cm high 45 m² Mixed grazed grasses <10 cm high 60 m² Mixed deciduous scrub <2 m high 70 m² Car park (tarmacadam) 60 m² etc.

The more detail being recorded the more survey effort that will be required to collect consistent information. For linear (terrestrial transect or stream) surveys, record habitat/vegetation type and height at regular points as suggested in Section 3.2.4 |

Variable	Units, etc.	Notes
Habitat connectivity	Presence of or distance to connecting features and/or barrier features, or quantified measures of these features at the site/within a certain distance; distance to other habitat patches/ ponds	e.g. presence of hedgerows, streams, minor roads, major highways, rivers (flowing water could be a connecting *or* barrier feature depending on the species and rates of flow!) or pond 120 m from major highway pond 40 m from hedgerow 3 other ponds within 500 m etc. or 125 m of mixed broadleaved hedgerow within 500 m of pond 56 m of minor road within survey site boundary 810 m of major highway within site boundary etc.
Threats	Presence or quantified impact of known/obvious threats	e.g. abundant predators (fish, waterfowl) invasive exotic species pollution (evidence of chemical use or chemical analyses) etc.
Site status	Threatened or protected (if known)	e.g. zoned for development substantial visitor impact – boating within a National Park protected because of the presence of rare species local protection only etc.
Site management	Management activities, plans and rationale where known	e.g. streambed dredged annually woodland on 10-year coppice rotation light grazing in summer subject to management plan for protected species etc.

Table 3.2 (*continued*)

Variable	Units, etc.	Notes
Water quality	Through plant and/ or invertebrate species, or see Section 3.3.1 for assessing water chemistry	e.g. % cover in pond of submerged and emergent vegetation presence of particular aquatic plants presence of invertebrate groups known to thrive only in the cleanest water presence of invertebrates known to tolerate anoxic conditions etc.
HSIs	May include estimation of many of the variables suggested above, and provide a standard protocol for recording them. Results in an index value that can be used for direct comparisons	See below

3.3.3 Habitat Suitability Indices (HSIs)

HSIs can be a more rapid and objective method of generating habitat comparisons; usually a single index value is generated that can be used to compare different sites or changes at the same site over time. The HSI for great crested newt ponds and habitat was published by Oldham *et al.* (2000) and uses 10 factors, each being scored between 0.1 and 1 to generate the index value. Factors are *never* scored 0, however bad they may be, as this stops the index working! The 10 factors are listed below with notes on each one, but this advice should be used in conjunction with reference to Oldham *et al.*'s (2000) paper and/or the advice sheet available from ARG-UK (http://www.arguk.org/), where a map and graphs are given to aid the scoring process.

1. Geographic location – scored from a map based on newt distribution, easy enough unless you're very near a border between the zones.
2. Pond area – measure or estimate as accurately as possible in m².
3. Years of drought per decade – can be tricky if it's not obvious whether or not a pond dries up: ask a landowner or local if possible!
4. Water quality – easy to misinterpret! Scores better when aquatic invertebrates are diverse and abundant and aquatic plants are present. Score should be reduced if pollution is evident but *not* because of poor water clarity (visibility).

5. Percentage of shoreline shaded – scores '1' up to 60% and fairly easy to estimate percentages above this with practice (not to be estimated at night!).
6. Waterfowl impact – best estimated using the ARG-UK simplified system (1 = none present, 0.67 = present but little impact evident, 0.1 = severe impact).
7. Fish impact – a simple system is again best (1 = absent, 0.67 = possibly present but no evidence of impact, 0.33 = small numbers of e.g. stickleback present, 0.1 = large fish population present).
8. Pond count – the number of other ponds (not counting the focus pond) present within a radius of 1 km of the survey pond; divide the count by 3.142 to calculate pond density and score from the graph.
9. Terrestrial habitat – can be one of the more tricky ones to score (see below), especially for anyone new to amphibian surveys; the presence of natural vegetation such as rough grassland or broadleaved woodland, as well as shelter (e.g. logs, rocks, mammal burrows) increases the value of the habitat for newts (see also below).
10. Percentage of macrophyte (aquatic plant) cover – includes submerged and emergent plants in the pond – but *not* duckweed (*Lemna* spp.). Comes with practice: ARG-UK provide a diagram to help estimate the percentage cover (and see below).

Some of these factors are simple parameters to be measured and others (such as pond area and aquatic plant coverage) cause some surveyors more difficulty. If you want to use (or even develop) HSIs, *get some practice*! Take some friends to a pond and get each to estimate, anonymously, the pond's area, then write down and compare the values … next measure the pond area as accurately as you can and see who was closest; discuss how that person arrived at their estimate! I get students to do this exercise when running training courses and, after a few goes at it, even the worst show substantial improvement. You can of course do this for any factors that involve areas or proportions, such as cover and shading.

The great crested newt HSI also includes the scoring of available terrestrial habitat based on what that species likes, so for that factor some knowledge or experience is required. That also comes with practice, of course, and two ponds surrounded by very different habitat can be just as suitable, something you might only realize if you've done lots of surveys. It's useful to shadow someone whilst they're recording HSI factors if you can, then try a few scores on your own and see how close you are to your experienced friend!

The final HSI value is the geometric mean of the 10 scores (in other words the tenth root of all the scores multiplied together):

HSI = (factor 1 × factor 2 × factor 3 × factor 4 × factor 5 × factor 6 × factor 7 × factor 8 × factor 9 × factor 10)$^{1/10}$

You can set up an Excel spreadsheet to work out the final index value for you. This will always be between 0 and 1, with ponds most suitable for great crested newts scoring nearer to 1:

- HSI <0.5 means poor suitability
- HSI in the range 0.51 to 0.69 means below average to average suitability
- HSI >0.7 means above average
- HIS >0.8 means excellent suitability.

The advantage of using an HSI is that you're told exactly what to measure and record, and how to score your measurements to arrive at the index value. I've found that, with practice, HSI factors can be assessed and recorded reasonably quickly, which is one reason this particular system is now very widely used in the UK by both conservationists and ecological consultants. Note that the index value is a guide to how suitable that habitat is for the species the HSI covers and *is not* necessarily an indicator of the size of the population present! Remember that amphibian populations can fluctuate substantially and anyway any changes in population size will probably track habitat changes over time (e.g. positive habitat management might result in a larger population some years later). You can surmise that areas with better habitat may support more amphibians and you could correlate lower population sizes with changes in HSI at the same site over time, but your HSI isn't a substitute for population surveys! I've based this advice on HSIs on my experiences with great crested newt surveys in Britain, but common toad and reptile HSIs are also being developed in the UK. See also related work such as that on modelling habitat requirements for western toads (*Anaxyrus boreas*) in Idaho, USA, by Bartelt *et al.* (2010) and for amphibians in deserts by Dayton and Fitzgerald (2006), if HSIs are of particular interest to you.

Finally, though HSIs are designed specifically to provide comparative data on their target species, they can be used *with caution* as a general means of comparing habitats. So NARRS surveyors in the UK are asked to carry out an HSI at every pond surveyed whether that survey is in an area where great crested newts are present or not. Future NARRS results should be able to detect any broad changes in pond habitat within those results but the changes will probably have affected the species present very differently. Palmate newts, for example, tend to fare relatively well in shady, acidic ponds that would score badly under the great crested newt HSI criteria, and toads can do extremely well in ponds containing large fish that would wipe out great crested newts! An HSI is more of a blunt instrument than a delicate assessment tool and it's up to the surveyors using it to present their HSI results in a relevant context, and without drawing unwarranted conclusions.

4. After your survey

Once all your fieldwork is over, there are still a few tasks remaining to complete your survey project. These include dealing with the data you have collected and setting out any conclusions you can draw in a report that aids understanding. Effective reporting is essential. It will maximize the usefulness of what you've found out for the conservationists, decision-makers and other surveyors who will want or need to read about your survey! This chapter includes further hints on handling data, analysis (refer also to Sections 2.3 and 2.4) and advice on formatting your survey report to increase its impact, as well as suggestions about the distribution of your report.

4.1 Arranging your data for analysis

The most essential job, right after finishing fieldwork, is to audit the data you've collected. It's best, where possible, to deal with the data you're collecting as you go along, filing survey sheets, etc., carefully when you have time between fieldwork sessions. Of course, this isn't always possible due to fieldwork schedules, as well as any other demands on your time. After fieldwork you should check you actually *have* all your site visit forms, survey sheets, etc., have obtained any supplementary (e.g. meteorological) data you need, and that the raw data (sheets, forms, etc.) are filed and arranged in a way you find logical so that you can refer back to them easily during report writing.

The best (and often simplest) ways of handling, analysing and displaying survey data will require having the data available electronically, so your next task will be inputting your results into a spreadsheet (or downloading and formatting it, if you've been using a tablet computer or similar device for data collection). Excel or similar spreadsheet software can be used effectively for this.

There are many ways of arranging survey data in a spreadsheet – and what seems logical to some may not be to others – but the best advice is to *be guided by the resources available to you*. If, for example, you'll be using SPSS statistical software to carry out complex analyses, this software will require the data to be arranged in a format that it can use to perform those analyses (and other input formats will produce meaningless results or simply errors). Even fairly simple analyses (e.g. from within Excel) need the data set out in a particular way to work correctly. So try to start by entering the data in the format you'll need, and be prepared to manipulate the format if you then decide to use different software! As a general rule, however, it's recommended that data are arranged in a spreadsheet by:

- using each *row* of the spreadsheet for one observation, record, sample or replicate
- using each *column* of the spreadsheet for a different variable or parameter.

This is best illustrated by the examples in Tables 4.1 and 4.2. Though at first this format may not be intuitive, it is essential for any complex analyses and becomes second nature after a while. The same kind of format is often used for uploading batches of data to databases.

In Table 4.1, four visits to the pond by two different recorders took place in March and April 2014. Day, month and year each have a column of their own so can be analysed as individual factors (and the columns can be easily concatenated for submitting biological records, etc.). Day can also be recorded as 'day of year' (i.e. 1st January = day 1, 1st February = day 32, and so on) for other analyses. Most importantly, each count of each species has a *row* of its own. Absences (where no animals were seen on that visit) are also recorded and, if comparing these survey visits with data from other ponds where different species were found, it would also be necessary to insert rows showing an absence of, for example, common frogs for each visit. Recording absences in surveys (see Section 2.4) is also valuable when submitting survey data to national datasets (such as, in the UK, the National Biodiversity Network Gateway, https://data.nbn.org.uk/). The data fulfil the requirements for biological records because they include *who*, *where*, *when* and *what* as well as (for this survey) *how many*. This simple table shows that no toads were recorded in the pond after March and that palmate newts were generally more abundant than smooth newts across all the surveys.

For exploration of relationships within your data, you can use the functions of your spreadsheet software to pull out and arrange sections of it for specific tests, or if a different format is needed. At the same time, it's always wise to make a copy of the master spreadsheet just in case your rearrangements mess things up. I'll freely admit that this happens to me all the time and I'm often very glad to have copies of my master spreadsheet to fall back on!

Some of the data from a fictitious survey of Oakwood Park Pond (see also Section 2.3) are shown in Table 4.2. The same data are used to illustrate Section 4.2. Variables not being used for a particular test (e.g. in this case 'surveyor' and 'site') can be omitted. Though this format is best for analyses, see Section 4.2.6 for ways these data could be presented in your survey report.

Table 4.1 Data showing adult amphibians recorded over four visits to Dewlands Common Pond in 2014.

Recorder	Site	Day	Month	Year	Species	Count
JWW	DCP	12	03	2014	toad	47
JWW	DCP	12	03	2014	smooth newt	12
JWW	DCP	12	03	2014	palmate newt	22
JWW	DCP	23	03	2014	toad	3
JWW	DCP	23	03	2014	smooth newt	14
JWW	DCP	23	03	2014	palmate newt	28
AA	DCP	10	04	2014	toad	0
AA	DCP	10	04	2014	smooth newt	13
AA	DCP	10	04	2014	palmate newt	23
JWW	DCP	22	04	2014	toad	0
JWW	DCP	22	04	2014	smooth newt	8
JWW	DCP	22	04	2014	palmate newt	24

Table 4.2 Some of the data from a fictitious amphibian survey of Oakwood Park Pond.

Month	Species	Count	Rainfall (mm in previous 48 h)
March	great crested newt	12	29
March	great crested newt	8	33
March	great crested newt	16	35
March	great crested newt	19	39
April	great crested newt	32	20
April	great crested newt	23	23
April	great crested newt	45	44
April	great crested newt	22	18
April	great crested newt	38	29

It would be hard for me to go through all the possible statistical analyses that can be carried out on survey data here. Fortunately (for all of us!) some biologists are very good at statistics and have shared their experiences and advice on analysing data in a series of books devoted to the subject. What book(s) to use for guidance can in itself be a minefield (because there are many) but I find that I refer regularly to just a few. These include Dytham (2003), Hawkins (2005) and Gardener (2012). In particular, Gardener (2012) goes through examples of statistical analyses using Excel and the increasingly popular R software, which is freely downloadable, and Dytham (2003) includes examples using Excel, SPSS and MINITAB. Beebee (2013) also gives some very good advice on hypothesis testing and statistics in relation to herpetological studies.

So if you're using statistics to explore your survey findings, find a book (or books) you get on with and refer to it (or them) often. I rarely write anything involving statistical analysis that doesn't involve me checking my approach by reference to one of my favourites. It's a good idea to 'try before you buy' and borrow from a friend or library first, to avoid buying a text you may find unintuitive (and, let's face it, some of the stats books out there aren't very easy to grasp!). You should also check out published surveys that have used statistics to investigate data like your own: which tests have they used and why those particular ones? Advice on finding useful published examples is given in Section 2.4. It's also important to repeat that you *really should plan* your analyses *before* collecting any data, otherwise you could find that your results won't answer the questions you set out to investigate! (Tip: *Wikipedia* has a whole host of pages covering data and statistical tests: these can be extremely informative, though the definitions get rather technical!)

4.2 Setting out your survey report

How long and detailed your survey report becomes depends not least on how much data you collected and how much analysis you want to present. Nevertheless, whether simply presenting your results for dissemination or preparing a scientific paper for publication in a peer-reviewed journal, all types of survey report should include four basic sections – *Introduction*, *Methods*, *Results* and *Discussion* – plus of course a good title. Setting out your report in these sections can be tricky if you aren't used to it but, in my experience, it *really helps* to use this format if you want your findings to be understood and useful. Some well-respected scientists still seem to get it a little wrong occasionally (or so it seems to me), so if you do find it tricky, you're in good company!

What information should go where in your report is discussed below but note that most journals, magazines, etc., have rules on *exactly* how your writing should be formatted and presented. That information is normally downloadable from their website or is printed in each copy of the publication. Possible publication outlets for *your* survey results can be found by tracking down published, peer-reviewed papers describing surveys like your own.

Start your report by creating a document containing all the headings you'll need, then populating each section with notes, and finally expanding the notes into understandable language. The vast majority of survey reports can be created using the following 11 headings.

4.2.1 Title

Your report's title should be descriptive of your survey without being too long or pedantic. It can be useful to write the title last of all, if it's a struggle to decide on one, but on the other hand a good title can help shape the contents of your report. *Survey Report* is of course inadequate but, for example, *Oakwood Park Amphibian Survey Results 2015* is just enough to describe what the document is actually about, for an unpublished report. The title can also provide more of a hint of what is covered in the report, where that's useful, for example *Amphibians Recorded during Oakwood Park Pond Survey 2015 and Relationships between Adult Newt Counts and Survey Conditions*. Check the titles of published surveys *you* find most informative for inspiration!

4.2.2 Other identifying information

- Surveyor/author names, addresses and e-mails; phone numbers are sometimes included if appropriate.
- Institutional affiliations (if not obvious from addresses).
- Date of report (for 'unpublished' reports).
- Sponsors/funders (in a journal paper these are listed in the *Acknowledgements*).
- How to reference the report (e.g. *Unpublished Report to Oakwood County Ranger Service*; journals have their own style for this).
- A table of contents and lists of tables and figures should be included for longer, less formal reports, as these help the reader to navigate; they're not usually provided for journal papers.

4.2.3 Abstract (or summary)

The *Abstract* is a potted summary of your survey; your question/s, what you did, what you found out and any conclusions. An *Abstract* is always included in journal papers and is useful for other kinds of report. It is best left until last to write! Don't be afraid to reveal some of your findings in your summary: that might be the piece of information that entices someone to read (and take notice of) the whole document. Abstracting summary information from your report gets easier with practice; you can get some practice by checking other *Abstracts* and seeing what they've included (and missed out). Most *Abstracts* are limited to a few hundred words. How would you summarize the example sections below (*Introduction*, *Methods*, *Results* and *Discussion*) in no more than 250 words?

4.2.4 Key words

Many journals allow authors to choose a number of key words (usually around eight) that will help other researchers searching for publications on a particular topic.

4.2.5 Introduction

The *Introduction* should contain the following elements:

- the background to the survey or study
- the specific survey question/s you're addressing
- site descriptions (and maps where useful)
- comparisons and references to previous studies that informed your survey project
- the importance of what you might find out
- other factors that informed your approach or that led to you carry out the survey.

The *Introduction* is where you explain what you're trying to achieve for the benefit of the reader. It sets out what to expect from the rest of the report (*without* starting to include any methods or results!). Be consistent in writing style throughout your report, always using the *past tense* when referring to your survey (because you're writing about something that has been completed) and, usually, in the *third person*. Contractions (e.g. *don't*, *wasn't*) should not be used. This book isn't about grammar so here's a short introduction to the Oakwood Park survey as an example.

Example Introduction

Four species of amphibian have previously been recorded from Oakwood County. The common frog (*Rana temporaria*) and common toad (*Bufo bufo*) are commonest in the north, and records of smooth and palmate newts (*Lissotrition vulgaris* and *L. helveticus*, respectively) occur throughout the county (see Smith, 2006). The great crested newt (*Triturus cristatus*) has not been recorded there since the late 1800s (Oldguy, 1895); this species is, however, known from several sites in nearby Birchtown (Madeup, 2010), less than 4 km from Oakwood Park, and is widespread in Birchtown Borough (Madeup, 2010; Faker and Strange, 2013). No formal amphibian surveys of Oakwood Park have been carried out since the site was included in a county-wide atlas project in 2006 (Smith, 2006). Following sightings of unidentified large newts by visitors to Oakwood Park in 2014, an amphibian survey of Oakwood Park Pond was carried out to identify the species present there and inform future site management.

Oakwood Park is a 26-ha municipal park in the north-west of Oakwood County. It consists mainly of improved (short) and rough grassland, with areas of oak woodland and scrub. The Park's pond (ZY 1234 5678, 57 m.a.s.l.), see Example Figure 1 [*not shown in this example*], is ca. 1,450 m in area and was deepened (to max. 2.5 m) following designation of the Park in 2003. Shallower areas (ca. 0.3 m deep) with aquatic vegetation (including *Potamogeton* sp.) occur on the pond's south side and areas of emergent vegetation are present. Drying occurs only in years of very high summer temperatures, such as in 2004 (Oakwood Ranger Service, 2013).

The numbers of visitors to Oakwood Park have been rising consistently since a recorded 34,000 in 2012 (Oakwood Ranger Service, 2013). This has resulted in increasing conflicts between visitor needs and efforts to conserve the Park's biodiversity. The presence of great crested newts in the Park would require the implementation of additional management prescriptions in light of increasing recreational use. This newt species is protected under national legislation and at European Union (EU) level, with statutory obligations to maintain and enhance its conservation status.

The same information could, of course, be written down in several different ways (or with more detail and background information) and you can swap the order of some of the facts around later on if another way seems better (I often do this!). If you're surveying and comparing a number of sites, describe each one in the same way (include a table of site characteristics if needed) and say how you selected them.

4.2.6 Methods (often called Materials and Methods)

The *Methods* section simply describes what you did in your survey, and nothing more. I really dislike vague descriptions of methods: *Forty amphibian breeding ponds in northern France were surveyed …* isn't very much use to anyone wanting to compare their results later on! It is better to assume that the reader knows nothing about the survey area and to tell them as much as you can, without excessive detail.

Example Methods

A total of 23 amphibian survey visits were carried out at Oakwood Park Pond during March and April 2015. Daytime visual surveys were conducted for anuran spawn counts and to establish the presence of newt eggs, and adult amphibians present were positively identified by capturing with a hand-held net. Formal amphibian counts (nine) were conducted by weekly night torching, throughout the survey period, using a one-million candlepower torch. Nets and other equipment were cleaned with veterinary disinfectant after each survey visit as they were also used for surveys elsewhere in Oakwood. Bottle trapping was not carried out because traps were destroyed during initial trials. Additional (rainfall) data were obtained from Oakwood County Meteorological Service.

Possible relationships between rainfall and newt counts, and between survey timing (month) and newt counts, were examined using Spearman's Rank Correlations and Kruskall-Wallis tests, respectively, to inform future surveys in the Park. Analyses were carried out with R statistical software (R Core Team, 2013).

Our fictitious study is rather limited in scope, just to keep things fairly simple! State all the survey methods used, in the order that you used them, and where appropriate say why you used those methods over others that could have been used. The software you used for analyses should also be stated and referenced. Some journal papers involve a lot of different analyses and will require several paragraphs for this, so that anyone wanting to repeat and compare your methods with their own can do so. As usual, published examples are the best way to check you're describing your survey methods and analyses in an appropriate way.

4.2.7 Results

Display your results in a way they can be understood easily: using tables and figures is often the best way to avoid needing several paragraphs of descriptive text and can be more effective. Nevertheless, tabular and graphical displays of your data need to be linked with at least some explanation.

Example Results

Spring 2015 amphibian surveys at Oakwood Park Pond detected the presence of common frogs, common toads, great crested newts and at least one species of smaller newt (*Lissotriton* sp.) through daytime visual searches for anuran spawn and newt eggs. This is the first record of great crested newts in Oakwood County since 1895 (see Oldguy, 1895). No palmate newts were detected by opportunistic netting of adult amphibians; however, smooth newt (*L. vulgaris*) presence was confirmed by multiple captures of adult males in breeding condition. These species can be distinguished using the diagnostic characters listed by Madeup (2010). Anuran spawn counts, adult newt counts and rainfall data are shown in Example Table 1.

Example Table 1 Anuran spawn and adult newt counts at Oakwood Park Pond in Spring 2015. Spawn counts are cumulative.

Date	Rainfall (mm in previous 48 h)	Common frog (total spawn clumps)	Common toad (total spawn strings)	Great crested newt count (adults)	Smooth newt count (adults)
5th March	29	5	0	12	4
12th March	33	7	0	8	16
19th March	35	1*	0	16	17
26th March	39	0	0	19	26
2nd April	20	0	0	32	23
9th April	23	0	7	23	22
16th April	44	0	32	45	20
23rd April	18	0	6*	22	19
30th April	29	0	0	38	21

* Unhatched spawn remaining on this date.

No toadspawn was recorded until 9th April (Example Table 1), though up to 18 adult toads were seen during night newt counts prior to that date (data not shown). Frogspawn had begun hatching by 19th March and toadspawn by 23rd April.

Great crested newt counts were not significantly related to rainfall in the previous 48 hours (Spearman's Rank Correlation; $S = 126.03$, rho $= -0.06$, $p = 0.90$ n.s.) but were significantly higher during April than in March (Kruskal–Wallis test; $KW = 6.05$, d.f. $= 1$, $p = 0.01*$), see Example Figure 2. Smooth newt counts were not significantly related to either rainfall or month.

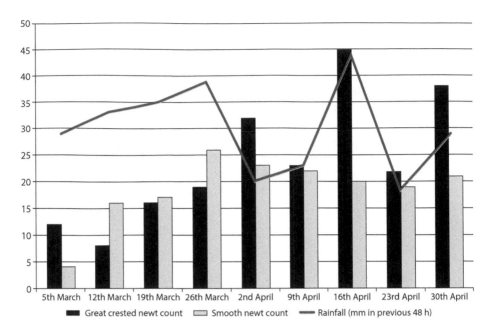

Example Figure 2 Counts of adult great crested and smooth newts in relation to rainfall at Oakwood Park Pond in Spring 2015 (John W. Wilkinson).

The text introduces the general findings of the survey but the actual survey data are presented in the table. Footnotes to a table (as used in Example Table 1) are often useful to explain anything that doesn't fit neatly inside it. Note that the results of statistical tests are described in a certain way for each test (depending on which ones were used), including whether or not the result was *significant* (see Box 4.1).

The table shows the relevant survey data but isn't particularly engaging for the reader. The figure displays the newt data much better. It isn't really necessary to display non-significant results graphically but the smooth newt and rainfall data fit easily onto the figure and the count differences are obvious and interesting. If you're displaying *all* your data graphically, there's no real need to use a table at all! For a more complex or detailed study, careful decisions have to be made as to how best to display the data, and which variables to include together on each figure. A simple legend should be included to explain each variable shown on your figures, and any tables and figures you use should *always be referred to* in your narrative text. A very useful section on displaying statistical results graphically is given by Gardener (2012).

Titles of tables go *above* and the titles of figures go *below*. I'm not certain why this happens but it is the convention irrespective of the publication concerned! Individual publications also have other rules on the layout of tables and figures, as well as on how they should be submitted, so do check these out before spending a lot of time on formatting. For an informal report, outputs like the Excel ones I've used here are usually acceptable.

BOX 4.1 SHOWING SIGNIFICANCE

The significance of a statistical test against a chosen threshold value – α (alpha), often α = 0.05 – is shown using p (= probability the result was due simply to chance or error); '$p = 0.02*$', for example (which is *below* the α threshold), shows that the result was significant at α = 0.05, indicated by the '*'. Other α values are sometimes used – for example α = 0.01 (**) or even α = 0.001 (***) – but there's usually a particular reason if and when they are. Use 'n.s.' when a result is *not* significant. Other test outputs (e.g. 'S' and 'rho' from the Spearman's Rank Correlation) are shown along with the p value but often mean very little to anybody but hard-core statisticians!

4.2.8 Discussion

This, of course, is where you discuss your findings and interpret how you think they're useful. The best way to open the *Discussion* is often a sentence or two summarizing the results you want to focus on. Next, compare what you found with results from other studies and add any reasons your results might *not* be as good as you'd hoped (someone else will probably point this out if you don't!). Unless you're simply making the results public, say why your survey findings were important and what the implications might be … and finally, discuss additional survey work or other studies suggested by your results.

Example Discussion

Four species of amphibian were recorded in Oakwood Park Pond in 2015 including, for the first time, the protected great crested newt. The species was not found there during surveys by Smith (2006). The survey revealed at least a 'medium' population of great crested newts (>10; see Example Table 1), however bottle trapping was not carried out in the present study and use of this method (if possible in the future) may result in higher counts. Dense aquatic vegetation currently makes thorough torching difficult in some areas of the pond.

Future management of Oakwood Park should take into account the aquatic and terrestrial needs of great crested newts, with visitor impacts (see Oakwood Ranger Service, 2013) directed away from key habitat features such as hibernacula. Management should be informed by ongoing newt surveys and, based on the torching results from this study, these should take place during April in order to offer the best chance of obtaining maximum counts. Earlier surveys will be necessary to monitor frog- and toadspawn numbers and trends if these are to be included in future monitoring efforts. A survey of terrestrial habitats to reveal the locations of amphibian hibernacula and main foraging areas would also be informative.

Monitoring of, and management for, great crested newts at Oakwood Park is particularly important in light of the fact that this is the only known site in Oakwood County for this species. Additional pond surveys between Oakwood Park and Birchtown should also be carried out to determine the presence of great crested newts in other ponds in the north-west of the county. See Faker and Strange (2013) for possible survey locations.

It can be good to end on a note of optimism or intrigue, in this case the suggestion that there could be more undiscovered populations nearby. Focus on the message you want to convey – the answer to your survey question (did you find out what you wanted to?) – but don't actually ignore other results or your report seems lopsided and un-objective. If, as is more often the case with real data, the messages are more complex and need more interpretation than in my limited example, careful thought will be needed to explain your key messages in the *Discussion*. What usually helps for me is to set the *Discussion* aside for a while, then read the report to that point with a fresh mind, making notes on what you want to include as you go. Have a look at the way other authors have set out and discussed their results too.

Some journals allow an additional section called *Conclusions* that follows the *Discussion*, but it's usually expected that your conclusions are set out and addressed within the *Discussion*. If you want a short, punchy section setting out conclusions at the end of an informal report (e.g. to aid speedy understanding by funders), that's usually fine.

4.2.9 Acknowledgements

An opportunity to thank everybody who helped you out! Best to start with the funders and maybe end with any people who showed especial interest; many authors also list their grant and/or licence numbers.

Example Acknowledgements

I thank Oakwood County Council and Friends of Oakwood Park for funding this study. Megan Jones and the Oakwood Ranger Service provided valuable assistance throughout, and Mrs Nettle of Oak Cottage provided much hospitality and local information. Surveys were carried out under Protected Species Survey Licence No. GCN1234.

4.2.10 References

Journals and other established publications have their own preferred (compulsory, in fact) format for references, based usually these days on the *Harvard* style. Though there are different interpretations of this style, journals especially will provide examples of different types of reference to guide you. The Harvard style includes citing publications in the text in the way I've been, for example '...as found by Smith (2006)' or '...has not previously been recorded (Faker and Strange, 2013)'. Where no compulsory style guide is provided, I prefer to use a fairly simple version of it.

For books and informal reports:

Madeup, R.U. (2010) *The Amphibians of Oakwood and Birchtown.* Treescape Publishing, Birchtown.

Smith, J. (2006) Oakwood County Amphibian Survey Report 2006. Unpublished Report to Oakwood County Council.

For journal papers and articles:

Oldguy, I.M.N. (1895) Musings on the fauna of Oakwood County and its environs. *Oakwood Science Society Bulletin* **105**: 232–245.

Faker, Z. and Strange, P.F. (2013) Aquatic biodiversity and pond loss in Birchwood Borough. *Fictitious Zoology* **22**: 45–67.

There are lots of trivial variations on this style used by individual publications and/or universities and colleges. Check if you need to use a particular style, but many more examples can be seen in the (real!) references for this book. References are listed in alphabetical order by the first author's last name; add letters to the year if more than one publication by an author or authors is from the same year, for example '2013a'. Some publications still use a superscript number (e.g. [23]) to cite each reference and then list them in the cited order, but this is less usual. Perhaps one day everyone will agree on just one style we can all learn and stick to: but not yet!

4.2.11 Appendices

Appendices are by no means required but can be useful for displaying large datasets, tables, survey forms, additional maps, etc., or substantial blocks of data (such as genetic data from a large series of samples) that might be important to the report but would make it disjointed if included within the main text. Each appendix is usually given a number or letter.

4.3 Who needs to see your data and read your report?

Even if your report is a literary masterpiece, the information within it needs to get to the right people for there to be a consequent scientific or conservation impact. A survey study published in a peer-reviewed journal will, once accepted and published, be in the public domain but an informal report usually won't be. In many cases, survey reports need at least a few copies actually printed out, bound and mailed to some of the people who need to see and refer back to it. Equally, however, many others will just need a pdf copy e-mailed to them or the raw data sent in spreadsheet form. It's good to plan (and keep a list of) who you'll be sending hard and electronic copies of your data and/or reports to as your project progresses. Additionally, your employer or institution, or a conservation body, etc., may want to make your report fully available via their website. The following list includes most of the places amphibian survey data and reports need to be sent to.

- Local/regional recording scheme or Environmental Records Centre (might need copies of your report *and* all your raw data in a formatted spreadsheet).
- National recording scheme.
- National biodiversity information web portal.
- Local conservation organization/s, including 'Friends of' groups for reserves, parks, etc.
- National amphibian (and reptile, usually) conservation charity.
- National and/or statutory conservation agencies.
- The licensing authority (data only, unless otherwise specified; usually as a condition of your licence).
- Landowners or tenants.
- Site managers, including, for example, rangers and the management authority.
- Anybody who helped with your fieldwork.
- Your employer, university or institution.
- Any organization or person providing cash funding.
- Anybody else providing in-kind support (e.g. lending equipment or giving you desk space).
- People who've shown particular interest (e.g. helpful locals).
- Local Authority ecologist and/or planning department.
- Your local library.

Several of these headings will, in many cases, refer to just one person or organization but the list is not exhaustive. Add in anyone else you think should see your report or, if you're not sure, simply ask them if they'd like a copy!

4.4   Taking amphibian studies further

If, like me, you find amphibian surveying fascinating and rewarding, you may be inspired to take your interest to another level. Throughout the preceding sections of the book, I've emphasized the importance of *robust and comparative* data, as well as of keeping in mind your *survey questions*: just what do you want to know? The species found in an area, trends in counts of breeding amphibians or how different species use a pond might be revealed by a well-planned amphibian survey but, equally, your survey results might spark a whole new set of questions that need investigation with additional techniques. Maximum counts of breeding newts, for example, might work well comparatively (between ponds or years), but how do those counts relate to the total number of newts actually using the pond? Some additional amphibian study methods, all involving an element of surveying, are given below. Many techniques can be applied to understanding species' ecology and informing their conservation.

4.4.1 Population studies

Differences in the detectability of amphibians at breeding sites, as well as other factors, make it hard to say definitively that a maximum count represents the total population size, even if you can reasonably compare counts and detect trends using data collected in a standardized way. To estimate the size of the breeding population, a variety of calculations based on your survey results can be carried out (using simple equations or software packages). To obtain the data necessary for these calculations, a capture–mark–recapture (CMR) study (also known as mark–release–recapture) is carried out. This means being able to recognize any animals you've captured before. Method options range from simple estimates based on two sets of captures, such as the Petersen method, through to multiple sample estimates using sophisticated software packages such as the program MARK (White and Burnham, 1999; http://www.phidot.org/software/mark/). See, for example, Heyer *at al.* (1994) or Southwood and Henderson (2000) for details of possible calculation methods and of the assumptions each one makes about CMR data.

   Captured amphibians can be recognized in two ways, either by giving each capture some sort of mark (see below) or by recording the animals' own existing patterns. Many amphibian species have ventral, dorsal or other patterning that lends itself well to individual identification. The archetypal example of this is the belly patterns of great crested newts, now often used in studies on population size, longevity, breeding pond use and a host of other investigations (e.g. Miaud *et al.*, 1993; Griffiths *et al.*, 2010; and Fig. 4.1). Belly patterns are recorded easily using a clear plastic box and some foam to restrain the newt gently while a digital photo is taken. Comparison can be made visually with previous digital images or there is now software available that will do it for you (e.g. I3S, available from http://www.reijns.com/i3s/about/I3S_Spot.html). See also Gamble *et al.* (2008) and Courtois *et al.* (2012) for CMR studies based on recognizing individual patterns of marbled salamanders and dart frogs, respectively.

Figure 4.1 A male great crested newt temporarily restrained in a clear plastic box (which formerly held a popular brand of chocolates) for photographing its belly pattern. This particular newt is known as 'Clint' and has been a regular visitor to this pond for many years! (David Sewell).

When your study amphibian is drably coloured or lacks obvious individuality in markings, a mark must be given to identify captures. Such marks are usually varied by date of capture. Many years ago I helped with a study on European common toads that used toe-clipping to identify animals previously caught (a particular clipped toe denoting the date of capture). This seemed effective at the time but nowadays we'd be much more concerned about the possible effects on the amphibians marked in this way, as well as whether or not the clip affects likelihood of recapture. Toe clipping is discussed by, for example, Golay and Durrer (1994) and Perry *et al.* (2011). Such an 'invasive' method of marking may also require licensing in many jurisdictions. There are, in any case, more acceptable techniques that are at least as effective. I used a Panjet dental inoculator to mark toads for my doctoral research (Wilkinson, 2007): this arcane device fires a blob of blue dye through an amphibian's skin under high pressure, without causing any kind of wound. Unfortunately, Panjets are now very hard to get hold of (and can be expensive even if you can!). The modern equivalent is to use colourful visible implant elastomer (VIE) tags, which can be effective on even small amphibians (though see Grant, 2008) and caecilians (Measey *et al.*, 2001). If budget is not a concern, passive integrated transponder (PIT) tags can be implanted under the skin, and each is completely unique, allowing true individual recognition for longer term population studies. The cost of PIT tags, as well as the required implanter and tag reader, is coming down all the time in real terms (see suppliers in Section 5.5) and this relatively straightforward marking method will no doubt see more applications in future studies.

4.4.2 Radio tracking

Recording comparative information, such as the habitats present where your study amphibians are found, is an important part of any survey. But how do those amphibians use each habitat? Pitfall trap and refugia studies (discussed in Section 3.2) can be illuminating but more detailed investigations are possible using techniques such as radio tracking. This, however, requires substantial resources: a receiver and tracking antenna can cost around £1000 (US$1500) and each transmitter around £100 (US$150), depending on the type used. External fixing of transmitters is also very difficult with amphibians and, though internal ones are available and can work well, implantation will normally require the services of a knowledgeable veterinarian. Licensing and ethical approval will also probably be required. Companies specializing in tracking technology (see Section 5.5) are generally helpful and can advise on the latest transmitters that might suit your study. These are now reducing in size and cost, making studies on amphibian range, migration and habitat use increasingly attractive. Successful investigations using radio tracking have been carried out by, for example,

Lamoureux and Madison (1999), who tracked green frogs (*Rana clamitans*) to determine their hibernation sites, Jehle and Arntzen (2000), who radio tracked newts following their breeding period, and Oromí *et al.* (2010), who investigated the ecology of natterjack toads.

4.4.3 Landscape-level (GIS) studies

No matter how good your survey is, estimates of how much habitat there is for a particular species in an entire region or country are rarely achievable without some kind of extrapolation. Quantification of this measure can be very useful for conservation status assessment (Section 2.1), especially if it is suspected that the amount of available habitat is diminishing. To categorize and measure all suitable habitats purely through field survey would require an immense amount of time and resources, and probably a vast army of dedicated surveyors!

The extent of suitable habitat for a species can, however, be modelled over large areas based on surveys of a sample of sites or presence records from within the area being considered. Software such as MaxEnt and Biomapper (e.g. Guisan and Zimmerman, 2000; Phillips *et al.*, 2006; Menéndez-Guerrero and Graham, 2013; French *et al.*, 2014), used within a GIS package in conjunction with land cover and environmental datasets, can interrogate the areas around species records and identify the characteristic variables found there. The software then looks for other areas in the landscape that match those characteristics and the quantification of suitable habitat becomes possible without blanket surveys. This process can also be used to target future surveys to areas that might support a species but from which no records exist.

Developments of GIS modelling include investigating barriers in the landscape (roads, rivers) and other areas that are inimical for many amphibians, such as large conurbations ('high-resistance areas'). In the same way, connectivity between populations can be examined by identifying areas that are hospitable to migration.

Many other spatial applications, such as trialling the landscape-level effects of new developments, are becoming possible as the techniques develop. A key element of using GIS-based habitat models effectively for conservation applications is to ensure the models are ground-truthed with additional surveys, and that still requires finding the resources for some fieldwork on which model improvements can be based. I've found, however, that the most important resource requirement for applying spatial modelling techniques to amphibian conservation is somebody who is very good indeed at using GIS!

4.4.4 Taxonomic studies

Every species of amphibian is unique but some are outwardly very similar! Similar-looking species may nevertheless have very different ecological needs that are important to understand if you want to conserve them, so there have to be means of telling them apart. Modern taxonomists use multiple techniques, the results of which need to agree before a species can be 'split' and accepted by the scientific community (see Box 1.1). These typically include morphometric and genetic comparisons. The understanding of different aspects of genetics, for example studies on variations in microsatellite DNA and/or mitochondrial DNA, is increasingly used to understand the relationships between similar species. Such studies have been carried out recently on several European amphibians (and many others) that were previously regarded as single species with a large geographic range (e.g. Recuero *et al.*, 2012; Arntzen *et al.*, 2013; Valbuena-Ureña *et al.*, 2014). Importantly, even highly laboratory-based taxonomic investigations rely on effective field surveys to locate, capture and measure or sample the study animals. Genetic samples are usually acquired through a small tissue sample or swab (see Fig. 2.4). If you're interested in developing skills in genetics, I can recommend the molecular ecology textbook by Beebee and Rowe (2007); see also Frankham *et al.* (2002) and Beebee (2005).

4.4.5 Studies on captive amphibians

This category covers a huge range of possible studies, many of which have the potential to generate information useful in species' conservation. Effective surveying to locate animals, as well as recording location and other data on specimens collected, is essential both to make sense of any results and for use in maintaining the captives ethically and in good health. Captive amphibian studies include (but are not limited to) courtship, mate selection, larval development, growth and feeding ecology.

Breeding amphibians in captivity can help our understanding of reproductive ecology and factors affecting spawn and tadpoles that can be difficult to study in the field (e.g. Weygoldt, 1980). Indeed, the eggs and tadpoles of some tropical amphibians are as yet undescribed. Captive amphibians are also playing a role as insurance colonies while wild populations of some species are critically threatened. This is particularly true of species affected by the chytrid fungus (see Section 2.6) and may become so for range-edge and upland populations being affected by climatic change (e.g. Pounds and Crump, 1994; Davidson *et al.*, 2002). Captive (*ex situ*) conservation efforts are coordinated internationally for the most endangered amphibians, and studbooks are maintained (see http://www.waza.org/en/site/conservation/international-studbooks).

If your surveying suggests further questions that might be answered by captive studies, carry out as much research as you can on the needs of your intended study species (or, if poorly known, their close relatives) and ensure you can access the resources to provide appropriate housing, temperature, lighting, humidity, nutrition, etc. You will also need to anticipate how long you might have to keep your captives and what will happen to them once your study is over. If, for example, they are housed in facilities with other species, is it appropriate for them to be released back into the wild? What is the risk of your study animals acquiring diseases from other captives or, indeed, transferring novel diseases to existing captive collections? These are real risks that zoos and hobbyists face regularly now that the origins of chytrid and other pathogens are becoming better understood (see Weldon *et al.*, 2004 for an overview; and Martel *et al.*, 2014 for worrying new information). Some published resources on keeping captive amphibians happy and healthy are given in Section 5.4 but, make no mistake, a serious study involving captive amphibians is a very considerable commitment. As with any new information that might help amphibian conservation, any results of captivity-based studies should be published and made widely available.

5. Resources to help you

5.1 Example survey forms

Casual amphibian records – always including *what*, *where*, *when* and *who*, of course – might only require jotting the data down in a field notebook along with any additional comments, but systematic surveys benefit from a well-designed form. Each survey is unique and survey questions differ in the details, so the examples given here are simply suggestions of how forms can be laid out. Often, your survey form is a compromise between including all the information you would like to record and the practicality of using it in the field: there's only room for so many boxes! An *Additional comments* box is useful on most forms for data you didn't anticipate you'd want to record, or for exceptional observations.

Figure 5.1 is an example of a form that can be used for recording site and habitat information. It's often best to record these data on a preliminary site visit when you won't be distracted by actual amphibian surveying, but remember to repeat the exercise if anything changes as your survey progresses. On your site sketch, you can also record trap locations, etc. (see Section 3.1.1). The data recorded in the form shown in Figure 5.2 could, with amendments, go on the reverse of a site information form and other boxes could be added for recording water chemistry, or indeed any data you want to collect (see Section 3.3). An audio survey of calling anurans, for example, will need a box to record an index value for the number of males calling (see Section 3.2.6). Just be careful your forms don't get too 'busy' or you'll quickly tire of filling them in!

Figure 5.3 shows a recent version of the form given to surveyors taking part in NARRS in the UK. Habitat data (in this case a simplified version of the great crested newt HSI) and species data are recorded on the same form and there's space for data from up to four visits, as well as data on variations in survey conditions. Again, each recording or monitoring scheme has its own requirements and will therefore employ a bespoke recording form.

SITE SKETCH. Grid spacing = _____m between lines.

Site name/location				
Grid reference/projection		Elevation m.a.s.l.	Date	
Surveyor/s				
Surveyor contact				
Landowner/s				
Landowner contact				
Designation/s?		Habitat type (within ___m of _____)		% cover
Pond size/stream length				
Max. depth	Natural?			
Substrate	Shading %			
Emergent veg. spp. / %	Aquatic veg. spp. / %			
Additional comments				

Figure 5.1 Example site information form (John W. Wilkinson).

Recorder/s			Recorder contact						
Date	Site name/location					% surveyed			
Start time	Grid reference/projection						Elevation m.a.s.l.		
End time	Weather now			Prior?			Wind		
Air temp.	Humidity			Cloud			Sun/moon		
Additional comments									

No.	Species	Stage/sex	SVL	W	Photo	Location/substrate	GPS	Activity	Time
1									
2									
3									
4									
5									
6									
7									
8									
9									
10									
11									
12									
13									
14									
15									
16									
17									
18									
19									
20									

SVL – snout-vent length (mm), W – weight (mass in grams)

Figure 5.2 Example recording sheet (individual amphibians) (John W. Wilkinson).

 **NARRS** National Amphibian and Reptile Recording Scheme

Amphibian
Survey Form 20__

 **amphibian and reptile** conservation

V. 2013

Data protection and copyright agreement

I understand that the information that I provide on this form, including my name and contact details, and those of any landowner, will be entered onto a computer database. I agree to share any intellectual property rights that may pertain to the data submitted on this form. Signature _____

Your details

Surveyor (Print name)

Address

Phone number

Email

Post code

Can we contact you if necessary? Yes / No

Landowner details *If the pond is on private land you must have the landowner/manager's permission to visit the site. If the pond is in a public access area it is still useful to know who owns the land.*

Name   Phone number

Address

Post code

Is the landowner/manager willing to be contacted if any follow-up is required?   Yes / No

Pond details

Pond grid reference   Nearest town

The grid reference needs to be in the form SP123456, or more detailed. For information on how to use the national grid please see www.ordnancesurvey.co.uk/oswebsite/freefun/nationalgrid/nghelp1.html

Pond name/address/reference number (and source)

If the pond no longer exists please tick here

Habitat suitability factors (refer to *Guidance Notes*)

1. Map Location. Score: A (optimal), B (marginal) or C (unsuitable).	**Optional**	6. Waterfowl impact. Score: 1 = major, 2 = minor, 3 = none.
2. Pond area in m². Estimate.		7. Fish presence. Score: 1 = major, 2 = minor, 3 = possible, 4 = absent.
3. Number of years in ten pond dries up. Estimate or ask landowner.		8. Number of ponds within 1 km (1: 25 0000 maps) not separated by barriers to dispersal.	**Optional**
4. Water quality. Score: 1 = bad, 2 = poor, 3 = moderate, 4 = good.		9. Terrestrial habitat. Score: 1 = none, 2 = poor, 3 = moderate, 4 = good.
5. Percentage perimeter shaded (to at least 1 m from shore). Estimate.		10. Percentage of pond surface occupied by aquatic vegetation (March–May). Estimate.

Water quality Bad = clearly polluted, only pollution-tolerant invertebrates, no submerged plants; Poor = low invertebrate diversity, few submerged plants; Moderate = moderate invertebrate diversity; Good = abundant and diverse invertebrate community.

Waterfowl impact Major = severe impact of waterfowl i.e. little or no evidence of submerged plants, water turbid, pond banks showing patches where vegetation removed, evidence of provisioning waterfowl; Minor = waterfowl present, but little indication of impact on pond vegetation, pond still supports submerged plants and banks are not denuded of vegetation; None = no evidence of waterfowl impact (moorhens may be present).

Fish presence Major = dense populations of fish known to be present; Minor = small numbers of crucian carp, goldfish or stickleback known to be present; Possible = no evidence of fish, but local conditions suggest that they may be present; Absent = no records of fish stocking and no fish revealed during survey(s).

Terrestrial habitat None = clearly no suitable habitat within immediate pond locale; Poor = habitat with poor structure that offers limited opportunities for foraging and shelter (e.g. amenity grassland); Moderate = offers opportunities for foraging and shelter, but may not be extensive; Good = extensive habitat that offers good opportunities for foraging and shelter completely surrounds pond e.g. rough grassland, scrub or woodland.

To complete the form overleaf please:
- Indicate detection of eggs and larvae with a tick (or question mark, if uncertain of species).
- Give counts of adults, immatures and frogspawn clumps.
- Record conditions under which the survey was carried out. **If BOTTLE TRAPPING, record the conditions at the time the traps were set out.**

SURVEY VISITS: Please complete up to **FOUR** visits to the pond nearest the lower left of your NARRS Square (note that information supplied from fewer visits **is still useful**). Please fill in the method **(VISUAL SEARCHING, NETTING, TORCHLIGHT SURVEY or BOTTLE TRAPPING)** used on each visit, and the survey details in the boxes below. Use bottle traps **ONLY** if trained, licensed and confident to do so. **Your survey results are valuable however many methods you can use!**
PLEASE ALSO RECORD ANY DEAD OR SICK AMPHIBIANS YOU SEE.

VISIT 1 – METHOD/S USED:					Date	
	Adu	Imm	Larva	Egg	Time	to:
Common frog					Air temperature °C	
Common toad					Water temperature °C	
Great crested newt					Water clarity (score 1–3)	
Palmate newt					Rain (score 0, 1, 2, 3)	
Smooth newt					Wind disturbing water (tick)	
Other species					Bright moonlight (tick)	
					% Shoreline surveyed	%
					Number of traps used	

VISIT 2 – METHOD/S USED:					Date	
	Adu	Imm	Larva	Egg	Time	to:
Common frog					Air temperature °C	
Common toad					Water temperature °C	
Great crested newt					Water clarity (score 1–3)	
Palmate newt					Rain (score 0, 1, 2, 3)	
Smooth newt					Wind disturbing water (tick)	
Other species					Bright moonlight (tick)	
					% Shoreline surveyed	%
					Number of traps used	

VISIT 3 – METHOD/S USED:					Date	
	Adu	Imm	Larva	Egg	Time	to:
Common frog					Air temperature °C	
Common toad					Water temperature °C	
Great crested newt					Water clarity (score 1–3)	
Palmate newt					Rain (score 0, 1, 2, 3)	
Smooth newt					Wind disturbing water (tick)	
Other species					Bright moonlight (tick)	
					% Shoreline surveyed	%
					Number of traps used	

VISIT 4 – METHOD/S USED:					Date	
	Adu	Imm	Larva	Egg	Time	to:
Common frog					Air temperature °C	
Common toad					Water temperature °C	
Great crested newt					Water clarity (score 1–3)	
Palmate newt					Rain (score 0, 1, 2, 3)	
Smooth newt					Wind disturbing water (tick)	
Other species					Bright moonlight (tick)	
					% Shoreline surveyed	%
					Number of traps used	

Water clarity 1 = good, pond bottom visible, 2 = intermediate, bottom visible in shallows, 3 = turbid, bottom not visible.
Rainfall 0 = none, 1 = yesterday, 2 = immediately prior, 3 = during survey.

Figure 5.3 Example recording scheme form (from UK NARRS).

5.2 Risk Assessments

Levels of risk for any activity can be assessed by scoring them and multiplying the *likelihood* of them happening by the *severity* of the outcome if they happen. Organizations often have their own ways of doing this (and their own scoring systems) but the scores and examples given here can be used as guidance. See also Section 2.6.

How likely is the hazard?
Unlikely (scores 1)
Moderately likely (scores 2)
Very likely (scores 3)

How severe could the outcome be?
Mild – inconvenience or minor injury requiring attention (scores 1)
Moderate – injury or illness requiring prompt attention (scores 2)
Severe – urgent major or multiple injuries, risk of fatality (scores 3)

Hazards can be arranged and scored in a table like the one below (Table 5.1), noting who is at risk and how the risk will be controlled. Only a few examples are given here: make a table of your own with as many rows as you need and assess the risks you might face in your own survey. You may disagree with how I've scored the risks shown below: that's fine! Each survey should have its own bespoke assessment, so say how *you* assess the risks you might face and what control measures *you* will take. Overall risk is then a product of both the likelihood and possible severity in your particular survey circumstances.

Often, your employer (or whomever you are surveying for) will require you to report certain hazards (see Table 5.2), if encountered (e.g. if any injury needs medical treatment), in an accident book or on a reporting form, so that risks can be reduced in the future. You should make sure you are aware of the circumstances under which this should happen. There may also be some assessed risks that are completely unacceptable: can you amend your survey to avoid these altogether?

Table 5.1 Some examples of assessed risks associated with amphibian survey activities.

Hazard	Risk	Who is at risk?	Likelihood score	Severity score	Overall risk	Control measures
Sharp objects (e.g. submerged in ponds, thorny vegetation)	Cuts, lacerations, infection	Surveyors, volunteer helpers	2	1	$2 \times 1 = 2$	Take appropriate care, carry a First Aid kit and disinfect cuts or skin punctures
Ponds or any deep water	Drowning	Surveyors, volunteer helpers	1	3	$1 \times 3 = 3$	Do not survey ponds alone, avoid slippery banks, wear footwear with good grip to minimize falling in
Uneven ground, ditches, hidden holes	Trips / falls leading to foot or other injuries	Surveyors, volunteer helpers	3	2	$3 \times 2 = 6$	Good footwear, visit survey sites in daylight prior to night surveys, avoid uneven ground with deep vegetation
Dogs and other large animals	Bites, injury, trampling	Surveyors, volunteer helpers	2	3	$2 \times 3 = 6$	Be aware of any animals around the survey site and avoid contact, leave site if necessary, seek medical attention for bites / injuries
Continue as needed …					See Table 5.2	

Table 5.2 Assessing overall level of risk, with examples of action needed. You must find out or decide what level is unacceptable and/or if you can bring in additional control measures to reduce high risks. Re-assess risks when extra controls are added.

		Likelihood (score)		
		Unlikely (1)	Moderately likely (2)	Very likely (3)
Severity (score)	Mild (1)	(1) Trivial – no action	(2) Low – control risk	(3) Medium – control now, report later
	Moderate (2)	(2) Low – control risk	(4) Medium – control now, report later	(6) High risk – control or STOP! Report immediately
	Severe (3)	(3) Medium – control now, report later	(6) High risk – control or STOP! Report immediately	(9) UNACCEPTABLE – STOP AND REPORT RIGHT AWAY!

5.3 Guides to amphibian identification and ecology

The following lists of books and other resources are not exhaustive but do represent those that I have personally found useful. New books and suppliers are appearing continually so a web search will often reveal fresh resources. These suggestions should nevertheless help you get started.

Africa
Baha el Din, S. (2006) *A Guide to the Reptiles and Amphibians of Egypt.* American University in Cairo Press, Cairo, Egypt.
Channing, A. (2001) *Amphibians of Central and Southern Africa.* Cornell University Press, Ithaca, NY, USA.

Channing, A. and Howell, K.M. (2006) *Amphibians of East Africa.* Edition Chimaira, Frankfurt am Main, Germany.

Du Preez, L. and Carruthers, V.C. (2009) *A Complete Guide to the Frogs of Southern Africa.* C. Struik Publishers, Cape Town, South Africa.

Glaw, F. and Vences, M. (2007) *A Field Guide to the Amphibians and Reptiles of Madagascar,* 3rd edn. Vences and Glaw Verlag, Cologne, Germany.

Schiøtz, A. (1999) *Treefrogs of Africa.* Edition Chimaira, Frankfurt am Main, Germany.

Sprawls, S., Howell, K.M. and Drewes, R.C. (2006) *Pocket Guide to the Reptiles and Amphibians of East Africa.* A&C Black, London, UK.

Asia

De Silva, A. (2009) *Amphibians of Sri Lanka.* Amphibian and Reptile Research Organization of Sri Lanka, Gampola, Sri Lanka.

Gardner, A.S. (2013) *The Amphibians and Reptiles of Oman and the UAE.* Edition Chimaira, Frankfurt am Main, Germany.

Goris, R.C. and Maeda, N. (2004) *Guide to the Amphibians and Reptiles of Japan.* Krieger Publishing Co., Malabar, FL, USA.

Inger, R.F. and Stuebing, R.B. (2005) *A Field Guide to the Frogs of Borneo.* Natural History Publications Borneo, Kota Kinabalu, Malaysia.

Iskandar, D.T. (1998) *The Amphibians of Java and Bali.* Pusat Penelitian dan Pengembangan Biologi, Jakarta, Indonesia.

Ranjit Daniels, R.J. (2005) *Amphibians of Peninsular India.* Orient Blackswan, Hyderabad, India.

Shrestha, T.K. (2001) *Herpetology of Nepal: A Field Guide to Amphibians and Reptiles of Trans-Himalayan Region of Asia.* Stephen Simpson, Norwich, UK.

Australasia/Oceania

Anstis, M. (2014) *Tadpoles and Frogs of Australia.* New Holland Publishers, Chatswood, NSW, Australia.

Jewell, T. (2008) *A Photographic Guide to Reptiles and Amphibians of New Zealand.* New Holland Publishers, Chatswood, NSW, Australia.

Menzies, J. (2006) *The Frogs of New Guinea and the Solomon Islands.* Pensoft Publishers, Sofia, Bulgaria.

Zug, G.R. (2013) *Reptiles and Amphibians of the Pacific Islands.* University of California Press, Oakland, CA, USA.

Europe

Arnold, N., Burton, J.A. and Ovenden, D. (1978) *Field Guide to the Reptiles and Amphibians of Britain and Europe.* Collins, London, UK (reprinted many times and still available).

Beebee, T.J.C. and Griffiths, R.A. (2000) *Amphibians and Reptiles.* New Naturalist 87. HarperCollins, London, UK.

Duguet, R., Melki, F. and ACEMAV (2005) *Les Amphibiens De France, Belgique et Luxembourg.* Biotope, Meze, France (in French).

Inns, H. (2009) *Britain's Reptiles and Amphibians.* WildGuides, Old Basing, UK.

Kuzmin, S.L. (1999) *The Amphibians of the Former Soviet Union.* Pensoft Publishers, Sofia, Bulgaria.

Sparreboom, M. (2014) *Salamanders of the Old World: The Salamanders of Europe, Asia and Northern Africa.* KNNV Uitgeverij, Zeist, The Netherlands.

Valakos, E.D., Pafilis, P., Sotiropoulos, K., Lymberakis, P. and Maragou, P. (2008) *The Amphibians and Reptiles of Greece.* Edition Chimaira, Frankfurt am Main, Germany.

North, Central and South America

Bartlett, R.D. and Bartlett, P.P. (2003) *Reptiles and Amphibians of the Amazon: An Ecotourist's Guide.* University Press of Florida, Gainesville, FL, USA.

Conant, R. and Collins, J.T. (1998) *Field Guide to Reptiles and Amphibians, Eastern and Central North America,* 4th edn. Houghton Mifflin Harcourt, Boston and New York, USA.

Corkran, C. and Thoms, C. (1996) *Amphibians of Oregon, Washington and British Columbia.* Lone Pine Publishing, Vancouver, BC, Canada.

Dodd Jr., C.K. (2013) *Frogs of the United States and Canada.* John Hopkins University Press, Baltimore, MD, USA (2 volumes).

Elliot, L., Gerhardt, C. and Davidson, C. (2009) *The Frogs and Toads of North America: A Comprehensive Guide to their Identification, Behavior, and Calls.* Houghton Mifflin Harcourt, Boston and New York, USA (includes a CD of calls).

Lee, J.C. (2000) *A Field Guide to the Amphibians and Reptiles of the Maya World: The Lowlands of Mexico, Northern Guatemala, and Belize.* Cornell University Press, Ithaca, NY, USA.

McCranie, J.R. and Wilson, L.D. (2002) *The Amphibians of Honduras.* SSAR Publications, Ithaca, NY, USA.

Oubouter, P.E. and Jairam, R. (2012) *Amphibians of Suriname.* Brill, Leiden, The Netherlands.

Petranka, J.W. (2010) *Salamanders of the United States and Canada.* Smithsonian Books, Washington, DC, USA.

Savage, J.M. (2006) *The Amphibians and Reptiles of Costa Rica.* University of Chicago Press, Chicago, IL, USA.

Schwartz, A. and Henderson, R.W. (1991) *Amphibians and Reptiles of the West Indies: Descriptions, Distributions and Natural History.* University Press of Florida, Gainesville, FL, USA.

Stebbins, R.C. (2003) *Field Guide to Western Reptiles and Amphibians,* 3rd edn. Houghton Mifflin Harcourt, Boston and New York, USA.

5.4   Other useful textbooks

Beebee, T.J.C. (2013) *Amphibians and Reptiles.* Naturalists' Handbooks 31. Pelagic Publishing, Exeter, UK.

Beebee, T.J.C. and Rowe, G. (2007) *An Introduction to Molecular Ecology,* 2nd edn. Oxford University Press, Oxford, UK.

Bennett, D. (1999) *Expedition Field Techniques: Reptiles and Amphibians.* Geography Outdoors, London, UK.

Carroll, R. (2009) *The Rise of Amphibians: 365 Million Years of Evolution.* John Hopkins University Press, Baltimore, MD, USA.

Cloudsley-Thompson, J.L. (1999) *The Diversity of Amphibians and Reptiles: An Introduction.* Springer Publishing, New York, NY, USA.

Davies, G. (ed.) (2002) *African Forest Biodiversity: A Field Survey Manual for Vertebrates.* Earthwatch Europe, UK.

Dytham, C. (2003) *Choosing and Using Statistics: A Biologist's Guide,* 2nd edn. Blackwell Science, Oxford, UK.

Gardener, M. (2012) *Statistics for Ecologists Using R and Excel: Data Collection, Exploration, Analysis and Presentation.* Pelagic Publishing, Exeter, UK.

Gardener, M. (2015) *Managing Data Using Excel: Organizing, Summarizing and Visualizing Scientific Data.* Pelagic Publishing, Exeter, UK.

Halliday, T. and Adler, K. (2002) *The New Encyclopaedia of Reptiles and Amphibians.* Oxford University Press, Oxford, UK.

Hawkins, D. (2005) *Biomeasurement: Understanding, Analysing and Communicating Data in the Biosciences.* Oxford University Press, Oxford, UK.

Heyer, W.R., Donnelly, M.A., McDiarmid, R.W., Hayek, L.-A.C. and Foster, M.S. (eds) (1994) *Measuring and Monitoring Biological Diversity: Standard Methods for Amphibians.* Smithsonian Institution Press, Washington and London.

Mattison, C. (1993) *Keeping and Breeding Amphibians: Caecilians, Newts, Salamanders, Frogs and Toads.* Orion, London, UK.

Monlezum, M. (2012) *Keeping Reptiles and Amphibians in the Classroom.* 5th Corner Publishing, Jacksonville, FL, USA.

Murphy, J., Adler, K. and Collins, J.T. (1994) *Captive Management and Conservation of Amphibians and Reptiles.* SSAR Publications, Ithaca, NY, USA.

Southwood, T.R.E. and Henderson, P.A. (2000) *Ecological Methods,* 3rd edn. Blackwell Science, Oxford, UK.

Wells, K.D. (2007) *The Ecology and Behavior of Amphibians.* University of Chicago Press, Chicago, IL, USA.

Wright, K.M. and Whitaker, B.R. (2001) *Amphibian Medicine and Captive Husbandry.* Krieger Publishing Co., Malabar, FL, USA.

5.5 Equipment suppliers

http://www.acowildlife.us/ or http://www.aco.co.uk/index.php (fencing, etc.)

https://www.bioquip.com/ (large variety of supplies.)

http://www.biotrack.co.uk/ (radio tracking supplies and PIT tags)

http://clulite.cluson.co.uk/ (torches)

http://www.hach.com/ (a variety of water-testing equipment, including turbidity meters)

https://www.justgloves.co.uk/ (laboratory gloves for fieldwork use)

http://www.nhbs.com/ (books and survey equipment)

http://www.rfidsystems.co.uk/ (PIT tags)

http://www.trapperarne.com (crayfish traps)

http://www.unobv.com/ (tags, etc.)

http://www.wildcareshop.com/ (much survey equipment, including nets, callipers, thermometers, etc.)

http://www.wildlifefencing.co.uk/ (drift fences, etc.)

5.6 Amphibian study and conservation organizations and societies

American Society of Ichthyologists and Herpetologists (ASIH) http://www.asih.org/

Amphibian and Reptile Conservation (ARC) http://www.arc-trust.org (UK, Europe and UK Overseas Territories)

Amphibian and Reptile Groups of the UK http://www.arguk.org/ (UK amphibian and reptile volunteer network)

Amphibian Research Centre http://frogs.org.au/ (Research and conservation for Australian frogs)

Asociación Herpetológica Española (AHE) http://www.herpetologica.es/ (Spanish Herpetological Society)

British Herpetological Society (BHS) http://www.thebhs.org

Centre de Coordination pour la Protection des Amphibiens et Reptiles de Suisse (KARCH) http://www.karch.ch/ (amphibian and reptile conservation in Switzerland)

Deutsche Gesellschaft für Herpetologie und Terrarienkunde (DGHT) http://www.dght.de/ (German Herpetological Society)

The Herpetological Society of Ireland (HSI) http://thehsi.org/

International Herpetological Society (IHS) http://www.ihs-web.org.uk/ (promotes responsible captive care and research)

Partners in Amphibian and Reptile Conservation (PARC) http://www.parcplace.org/ (species and habitats conservation network)

Reptielen Amfibiien Vissen Onderzoek Nederland (RAVON) http://ravon.nl (Reptile, Amphibian and Fish Conservation Netherlands)

Sociedade Brasileira de Herpetologia http://www.sbherpetologia.org.br/ (Brazilian Herpetological Society)

Societas Europaea Herpetologica (SEH) http://www.seh-herpetology.org (The European Herpetological Society)

See http://www.seh-herpetology.org/links/european_herpetological_societies for a fuller list of national herpetological societies in Europe

Societas Hellenica Herpetologica http://www.elerpe.org/ (Greek Herpetological Society)

Society for the Study of Amphibians and Reptiles http://ssarherps.org/ (US society with wider membership)

References

Acevedo, M.A. and Villanueva-Rivera, L.J. (2006) Using automated digital recording systems as effective tools for the monitoring of birds and amphibians. *Wildlife Society Bulletin* **34(1):** 211–214.

Arntzen, J.W., Recuero, E., Canestrelli, D. and Martínez-Solano, I. (2013) How complex is the *Bufo bufo* species group? *Molecular Phylogenetics and Evolution* **69(3):** 1203–1208.

Baker, J. (2013) Effect of bait in funnel-trapping for great crested and smooth newts *Triturus cristatus* and *Lissotriton vulgaris*. *Herpetological Bulletin* **124:** 17–20.

Bartelt, P.E., Klaver, R.W. and Porter, W.P. (2010) Modeling amphibian energetics, habitat suitability, and movements of western toads, *Anaxyrus(=Bufo) boreas*, across present and future landscapes. *Ecological Modelling* **221(22):** 2675–2686.

Beckmann, C. and Göcking, C. (2012) Wie die Motte zum Licht? Ein Vergleich der Fängigkeit von beleuchteten und unbeleuchteten Wasserfallen bei Kamm-, Berg- und Teichmolch. *Zeitschrift für Feldherpetologie* **19:** 67–78.

Beebee, T.J.C. (2005) Amphibian conservation genetics. *Heredity* **95:** 423–427.

Beebee, T.J.C. (2013) *Amphibians and Reptiles*. Naturalists' Handbooks 31. Pelagic Publishing, Exeter, UK.

Beebee, T.J.C. and Buckley, J. (2014) Relating spawn counts to the dynamics of British natterjack toad *Bufo calamita* populations. *Herpetological Journal* **24:** 25–30.

Beebee, T.J.C. and Griffiths, R.A. (2005) The amphibian decline crisis: a watershed for conservation biology? *Biological Conservation* **125(3):** 271–285.

Beebee, T.J.C. and Rowe, G. (2007) *An Introduction to Molecular Ecology*, 2nd edn. Oxford University Press, Oxford, UK.

Beja, P. and Alcazar, R. (2003) Conservation of Mediterranean temporary ponds under agricultural intensification: an evaluation using amphibians. *Biological Conservation* **114(3):** 317–326.

Biggs, J., Ewald, N., Valentini, A., Gaboriaud, C., Dejean, T., Griffiths, R.A., Foster, J., Wilkinson, J.W., Arnell, A., Brotherton, P., Williams, P. and Dunn, F. (2014) Using eDNA to develop a national citizen science-based monitoring programme for the great crested newt (*Triturus cristatus*). *Biological Conservation* **183:** 19–28.

Blaustein, A.R., Han, B.A., Relyea, R.A., Johnson, P.T., Buck, J.C., Gervasi, S.S. and Kats, L.B. (2011) The complexity of amphibian population declines: understanding the role of cofactors in driving amphibian losses. *Annals of the New York Academy of Sciences* **1223(1):** 108–119.

Boughton, R.G., Staiger, J. and Franz, R. (2000) Use of PVC pipe refugia as a sampling technique for hylid treefrogs. *The American Midland Naturalist* **144(1):** 168–177.

Bridges, A.S. and Dorcas, M.E. (2000) Temporal variation in anuran calling behavior: implications for surveys and monitoring programs. *Copeia* **2000**: 587–592.

Carrier, J.-A. and Beebee, T.J.C. (2003) Recent, substantial, and unexplained declines of the common toad *Bufo bufo* in lowland England. *Biological Conservation* **111(3)**: 395–399.

Chandler Schmutzer, A., Gray, M.J., Burton, E.C. and Miller, D.L. (2008) Impacts of cattle on amphibian larvae and the aquatic environment. *Freshwater Biology* **53(12)**: 2613–2625.

Cloudsley-Thompson, J.L. (1999) *The Diversity of Amphibians and Reptiles: An Introduction.* Springer Publishing, New York, NY, USA.

Courtois, E.A., Pineau, K., Villette, B., Schmeller, D.S. and Gaucher, P. (2012) Population estimates of *Dendrobates tinctorius* (Anura: Dendrobatidae) at three sites in French Guiana and first record of chytrid infection. *Phyllomedusa* **11(1)**: 63–70.

Davidson, C., Shaffer, H.B. and Jennings, M.R. (2002) Spatial tests of the pesticide rift, habitat destruction, UV-B, and climate-change hypotheses for California amphibian declines. *Conservation Biology* **16(6)**: 1588–1601.

Dayton, G.H. and Fitzgerald, L.A. (2006) Habitat suitability models for desert amphibians. *Biological Conservation* **132(1)**: 40–49.

Dewsbury, D. (2011) An alternative method for surveying and catching newts. *In Practice* **71**: 37–40.

Drechsler, A., Bock, D., Ortmann, D. and Steinfartz, S. (2010) Ortmann's funnel trap – a highly efficient tool for monitoring amphibian species. *Herpetology Notes* **3**: 13–21.

Dytham, C. (2003) *Choosing and Using Statistics: A Biologist's Guide*, 2nd edn. Blackwell Science, Oxford, UK.

Fogarty, J.H. and Vilella, F.J. (2001) Evaluating methodologies to survey *Eleutherodactylus* frogs in montane forests of Puerto Rico. *Wildlife Society Bulletin* **29(3)**: 948–955.

Frankham, R., Briscoe, D.A. and Ballou, J.D. (2002) *Introduction to Conservation Genetics.* Cambridge University Press, Cambridge, UK.

French, G.C.A., Wilkinson, J.W., Fletcher, D.H. and Arnell, A.P. (2014) *Quantifying the Status of Great Crested Newts in Wales.* NRW Science Report Series, Report 31. NRW, Bangor, UK.

Frost, D.R., Grant, T., Faivovich, J., Bain, R.H., Haas, A., Haddad, C.F., De Sa, R.O., Channing, A., Wilkinson, M., Donnellan, S.C., Raxworthy, C.J., Campbell, J.A., Blotto, B.L., Moler, P., Drewes, R.C., Nussbaum, R.A., Lynch, J.D., Green, D.M. and Wheeler, W.C. (2006) The amphibian tree of life. *Bulletin of the American Museum of Natural History* **297**: 1–370.

Gamble, L., Ravela, S. and McGarigal, K. (2008) Multi-scale features for identifying individuals in large biological databases: an application of pattern recognition technology to the marbled salamander *Ambystoma opacum*. *Journal of Applied Ecology* **45(1)**: 170–180.

Gardener, M. (2012) *Statistics for Ecologists Using R and Excel: Data Collection, Exploration, Analysis and Presentation.* Pelagic Publishing, Exeter, UK.

Gledhill, D.G. and James, P. (2008) Rethinking urban blue spaces from a landscape perspective: species, scale and the human element. *Salzburger Geographische Arbeiten* **42**: 151–164.

Golay, N. and Durrer, H. (1994) Inflammation due to toe-clipping in natterjack toads (*Bufo calamita*). *Amphibia–Reptilia* **15(1)**: 81–83.

Grant, E.H.C. (2008) Visual implant elastomer mark retention through metamorphosis in amphibian larvae. *The Journal of Wildlife Management* **72(5)**: 1247–1252.

Griffiths, R.A., Sewell, D. and McCrea, R.S. (2010) Dynamics of a declining amphibian metapopulation: survival, dispersal and the impact of climate. *Biological Conservation* **143(2)**: 485–491.

Guisan, A. and Zimmermann, N.E. (2000) Predictive habitat distribution models in ecology. *Ecological Modelling* **135**: 147–186.

Halliday, T. and Adler, K. (2002) *The New Encyclopaedia of Reptiles and Amphibians.* Oxford University Press, Oxford, UK.

Hamer, A.J. and McDonnell, M.J. (2008) Amphibian ecology and conservation in the urbanising world: a review. *Biological Conservation* **141(10)**: 2432–2449.

Hawkins, D. (2005) *Biomeasurement: Understanding, Analysing and Communicating Data in the Biosciences.* Oxford University Press, Oxford, UK.

Hayes, J.P., and Steidl, R.J. (1997) Statistical power analysis and amphibian population trends. *Conservation Biology* **11(1)**: 273–275.

Heyer, W.R., Donnelly, M.A., McDiarmid, R.W., Hayek, L.-A.C. and Foster, M.S. (eds) (1994) *Measuring and Monitoring Biological Diversity: Standard Methods for Amphibians.* Smithsonian Institution Press, Washington and London.

Jehle, R. and Arntzen, J.W. (2000) Post-breeding migrations of newts (*Triturus cristatus* and *T. marmoratus*) with contrasting ecological requirements. *Journal of Zoology* **251(3)**: 297–306.

Johnson, J.R., Knouft, J.H. and Semlitsch, R.D. (2007) Sex and seasonal differences in the spatial terrestrial distribution of gray treefrog (*Hyla versicolor*) populations. *Biological Conservation* **140(3)**: 250–258.

Kime, N.M., Turner, W.R. and Ryan, M. J. (2000) The transmission of advertisement calls in Central American frogs. *Behavioral Ecology* **11(1)**: 71–83.

Kröpfli, M., Heer, P. and Pellet, J. (2010) Cost-effectiveness of two monitoring strategies for the great crested newt (*Triturus cristatus*). *Amphibia–Reptilia* **31**: 403–410.

Kupfer, A., Müller, H., Antoniazzi, M.M., Jared, C., Greven, H., Nussbaum, R.A. and Wilkinson, M. (2006) Parental investment by skin feeding in a caecilian amphibian. *Nature* **440(7086)**: 926–929.

Lamoureux, V.S. and Madison, D. M. (1999) Overwintering habitats of radio-implanted green frogs, *Rana clamitans. Journal of Herpetology* **33(3)**: 430–435.

Lips, K.R., Reaser, J.K., Young, B.E. and Ibáñez, R. (2001) *Amphibian Monitoring in Latin America: A Protocol Manual.* SSAR Herpetological Circular No. 30. Available at: http:// ssarherps.org/publications/books-pamphlets/herpetological-circulars/.

Louette, G., Devisscher, S. and Adriaens, T. (2013) Control of invasive American bullfrog *Lithobates catesbeianus* in small shallow water bodies. *European Journal of Wildlife Research* **59(1)**: 105–114.

Madden, N., and Jehle, R. (2013) Farewell to the bottle trap? An evaluation of aquatic funnel traps for great crested newt surveys (*Triturus cristatus*). *Herpetological Journal* **23(4)**: 241–244.

Martel, A., Blooi, M., Adriaensen, C., Van Rooij, P., Beukema, W., Fisher, M.C., Farrer, R.A., Schmidt, B.R., Tobler, U., Goka, K., Lips, K.R., Muletz, C., Zamudio, K.R., Bosch, J., Lötters, S., Wombwell, E., Garner, T.W.J., Cunningham, A.A., Spitzen-van der Sluijs, A., Salvidio, S., Ducatelle, R., Nishikawa, K., Nguyen, T.T., Kolby, J.E., Van Bocxlaer, I., Bossuyt and Pasmans, F. (2014) Recent introduction of a chytrid fungus endangers Western Palearctic salamanders. *Science* **346(6209):** 630–631.

McCallum, M.L. (2007) Amphibian decline or extinction? Current declines dwarf background extinction rate. *Journal of Herpetology* **41(3):** 483–491.

Measey, G.J., Gower, D.J., Oommen, O.V. and Wilkinson, M. (2001) Permanent marking of a fossorial caecilian, *Gegeneophis ramaswamii* (Amphibia: Gymnophiona: Caeciliidae). *Journal of South Asian Natural History* **5:** 109–115.

Mendez, D., Webb, R., Berger, L. and Speare, R. (2008) Survival of the amphibian chytrid fungus *Batrachochytrium dendrobatidis* on bare hands and gloves: hygiene implications for amphibian handling. *Diseases of Aquatic Organisms* **82:** 97–104.

Menéndez-Guerrero, P.A. and Graham, C.H. (2013) Evaluating multiple causes of amphibian declines of Ecuador using geographical quantitative analyses. *Ecography* **36:** 1–14.

Miaud, C., Joly, P. and Castanet, J. (1993) Variation in age structures in a subdivided population of *Triturus cristatus*. *Canadian Journal of Zoology* **71(9):** 1874–1879.

Oldham, R.S., Keeble, J., Swan, M.J.S. and Jeffcote, M. (2000) Evaluating the suitability of habitat for the great crested newt (*Triturus cristatus*). *Herpetological Journal* **10(4):** 143–156.

Oromí, N., Sanuy, D. and Sinsch, U. (2010) Thermal ecology of natterjack toads (*Bufo calamita*) in a semiarid landscape. *Journal of Thermal Biology* **35(1):** 34–40.

Pechmann, J.H.K., Scott, D.E., Semlitsch, R.D., Caldwell, J.P., Vitt, L.J. and Gibbons, J.W. (1991) Declining amphibian populations: the problem of separating human impacts from natural fluctuations. *Science* **253:** 892–895.

Perry, G., Wallace, M.C., Perry, D., Curzer, H. and Muhlberger, P. (2011) Toe clipping of amphibians and reptiles: science, ethics, and the law 1. *Journal of Herpetology* **45(4):** 547–555.

Phillips, S.J., Anderson, R.P. and Schapire, R.E. (2006) Maximum entropy modeling of species geographic distributions. *Ecological Modelling* **190(3):** 231–259.

Platz, J.E. (1989) Speciation within the chorus frog *Pseudacris triseriata*: morphometric and mating call analysis of the boreal and western subspecies. *Copeia* **1989:** 704–702.

Pounds, J.A. and Crump, M.L. (1994) Amphibian declines and climate disturbance: the case of the golden toad and the harlequin frog. *Conservation Biology* **8(1):** 72–85.

Preston, D.L., Henderson, J.S. and Johnson, P.T. (2012) Community ecology of invasions: direct and indirect effects of multiple invasive species on aquatic communities. *Ecology* **93(6):** 1254–1261.

R Core Team (2013) *R: A Language and Environment for Statistical Computing*. R Foundation for Statistical Computing, Vienna, Austria. Available at: http://www.R-project.org/

Reading, C.J. (1998) The effect of winter temperatures on the timing of breeding activity in the common toad *Bufo bufo*. *Oecologia* **117(4):** 469–475.

Recuero, E., Canestrelli, D., Vörös, J., Szaby, K., Poyarkov, N.A., Arntzen, J.W., Crnobrnja-Isailovic, J., Kidov, A.A., Cogălniceanu, D., Caputo, F.P., Nascetti, G., Martínez-Solano,

I. (2012) Multilocus species tree analyses resolve the radiation of the widespread *Bufo bufo* species group (Anura, Bufonidae). *Molecular Phylogenetics and Evolution* **62**: 71–86.

Savage, R.M. (1961) *The Ecology and Life History of the Common Frog.* Pitman, London, UK.

Sewell, D., Beebee, T.J.C. and Griffiths, R.A. (2010) Optimising biodiversity assessments by volunteers: the application of occupancy modelling to large-scale amphibian surveys. *Biological Conservation* **143(9):** 2102–2110.

Southwood, T.R.E and Henderson, P.A. (2000) *Ecological Methods,* 3rd edn. Blackwell Science, Oxford, UK.

Snow, N.P. and Witmer, G.W. (2011) A field evaluation of a trap for invasive American bullfrogs. *Pacific Conservation Biology* **17(3):** 285–291.

Speybroeck, J., Beukema, W. and Crochet, P.A. (2010) A tentative species list of the European herpetofauna (Amphibia and Reptilia) – an update. *Zootaxa* **2492:** 1–27.

Stuart, S.N., Chanson, J.S., Cox, N.A., Young, B.E., Rodrigues, A.S., Fischman, D.L. and Waller, R.W. (2004) Status and trends of amphibian declines and extinctions worldwide. *Science* **306(5702):** 1783–1786.

Valbuena-Ureña, E., Steinfartz, S. and Carranza, S. (2014) Characterization of micro-satellite loci markers for the critically endangered Montseny brook newt (*Calotriton arnoldi*). *Conservation Genetics Resources* **6(2):** 263–265.

Weldon, C., Du Preez, L.H. Hyatt, A.D., Muller, R. and Speare, R. (2004) Origin of the amphibian chytrid fungus. *Emerging Infectious Diseases* **10(12):** 2100–2105.

Wells, K.D. (2007) *The Ecology and Behavior of Amphibians.* University of Chicago Press, Chicago, IL, USA.

Weygoldt, P. (1980) Complex brood care and reproductive behaviour in captive poison-arrow frogs, *Dendrobates pumilio* O. Schmidt. *Behavioral Ecology and Sociobiology* **7(4):** 329–332.

Wilkinson, J.W. (2007) *Ecology and conservation of the European common toad* (Bufo bufo*) in Jersey, British Channel Islands.* PhD thesis, University of Kent, Canterbury, UK.

White, G.C. and Burnham, K.P. (1999) Program MARK: survival estimation from populations of marked animals. *Bird Study* **46(S1):** S120–S139.

Glossary

Annuli the segmented rings found on the bodies of caecilians (singular: **annulus**)

Anura order of tail-less amphibians (frogs and toads)

Carboniferous Period geological period from about 360 million to about 300 million years ago

Caudata order of tailed amphibians (newts and salamanders)

Chytrid the pathogenic fungal organism *Batrachochytrium dendrobatidis*, responsible for many documented amphibian declines and some extinctions in various parts of the world; also now *Batrachochytrium salamandrivorans*, which is particularly harmful to caudates

Cloaca the genital and waste opening of amphibians

Epiphytic term used to describe a plant that grows upon another plant (often a tree) for support

Gymnophonia order of tropical, worm-like amphibians (caecilians)

Harvard style the style of parenthetical referencing used frequently (with minor variations) in scientific reports and journal papers

Hibernaculum a place where animals spend the winter (hibernate) (plural: **hibernacula**)

Hydroperiod the length of time a water body actually contains water

Metamorphosis the process by which a larval amphibian transforms into a juvenile resembling the adult

Microsatellite DNA short sections of variable DNA that can be used to examine relationships between organisms and in population studies

Mitochondrial DNA DNA found in the mitochondria of organisms' cells. It is inherited from the mother so can be used to investigate maternal lineages over time

Morphometrics the study of shape and form, sometimes used as part of investigations into amphibian taxonomy and evolution

Paedomorphic retaining larval traits (such as external gills) in breeding adults

Permian Period geological period from about 300 million to about 250 million years ago

SUDS sustainable urban drainage system – mitigation for drainage problems that may be caused by housing and other new developments, can be designed to incorporate amphibian habitats

Taxonomy the study of the description, classification and naming of organisms

Viviparous giving birth to young that have developed within the female (as opposed to **oviparous** – egg-laying)

Index

Page numbers in *italics* refer to figures and tables, and those in **bold** refer to Boxes.

CPSIA information can be obtained at www.ICGtesting.com
Printed in the USA
LVOW05*0851061015

457113LV00014B/26/P

9 781784 270049